Graphs as Mathematical Models

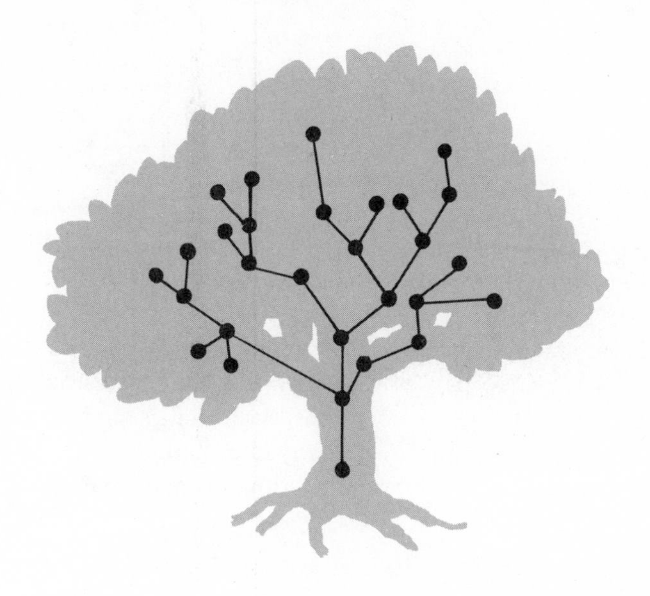

Gary Chartrand

Western Michigan University

Prindle, Weber & Schmidt, Incorporated
Boston, Massachusetts

To

Two Model Friends

Sue	*Curt*
Steffens	*Wall*

Library of Congress Cataloging in Publication Data

Chartrand, Gary.
Graphs as mathematical models.

Includes index.
1. Graph theory. I. Title.
QA166.C453 511′.5 77-5479

ISBN 0-87150-236-4
Printed in the United States of America.

Cover and text illustrations by Lois Orlando.
Technical art by Bill and Barbara Taylor.
Text design by David Chelton and the staff of Prindle, Weber & Schmidt.
Composed in Monophoto Times Roman by Composition House, Ltd.
Printed and bound by The Maple Press Company.

Preface

I have written this book with several objectives in mind:

> To teach the reader some of the topics in the youthful and exciting field of graph theory;
>
> To show how graphs are applicable to a wide variety of subjects, both within and outside mathematics;
>
> To increase the student's knowledge of, and facility with, mathematical proof; and last, but not least,
>
> To have some fun with mathematics.

Courses may be taught from this text that involve all four goals; other courses may minimize or eliminate the rigor of mathematical proof. Thus the text can be used for teaching students at all levels of undergraduate study.

Various versions of the notes which led to this book have been used for teaching courses at universities, colleges, and two-year colleges, the major differences being how the instructor emphasized or de-emphasized proofs. Each of these courses has included Chapters 1–3 and Section 4.1, with careful attention to Chapter 2.

The remaining material for these courses was chosen from the later chapters according to the instructor's tastes.

Recent years have seen increased demand for applications of mathematics. Graph theory has proven to be particularly useful to a large number of rather diverse fields. I have presented several problems throughout the text to illustrate various applications of graphs and graph theory. Appropriate graph theory concepts and results are introduced for the express purpose of "modeling" these problems mathematically. In the process, some of the theory of graphs is developed. The large variety of proofs used in this field can help strengthen the student's use of mathematical techniques.

Although graphs have numerous significant applications, the nature of the subject lends itself naturally to less serious uses. I have taken advantage of this to insert, now and then, a little humor into the discussion. It is my hope that I do not offend anyone with this use of mathematics.

If the book is to be used in a course which stresses mathematical proofs, then it might be wise for the student to read the Appendix, which discusses sets, relations, functions, wordings of theorems, and proof techniques. Exercises, sections, and chapters which involve a higher degree of mathematical content are starred and probably should be omitted if the course is to de-emphasize proofs. There are other exercises which require some mathematical arguments, and these should probably be omitted as well if the emphasis is strictly on concepts and applications.

Answers, hints, and solutions are provided to selected exercises. Some exercises have no specific answers and are intended as "discussion questions." Every chapter concludes with Suggestions for Further Reading, and I have briefly indicated the mathematical level of the references. The end of a proof and the end of the Preface are indicated by the symbol ▮. ▮

Acknowledgments

Appreciation is due to several people for their assistance, directly or indirectly, in the writing of this book.

The "flavor" of the book was influenced by my association and friendship with Jim Stewart of Lansing (Michigan) Community College. Shashi Kapoor taught the first course from the original notes, and his success encouraged me to continue writing. Mary Irvin assisted me a great deal with organizing the first draft of the manuscript. To all three of you, many thanks.

Several reviewers at various stages of the book's development provided valuable suggestions for improvement. It is a pleasure for me to acknowledge John Leonard, Linda Lesniak-Foster, Al Polimeni, Geert Prins, Sy Schuster, Don VanderJagt, and Curt Wall. In addition, I benefited from others who taught from the manuscript, namely Yousef Alavi, Brian Garman, John Roberts, and Jim Williamson.

A special note of thanks to my good friend Marilyn Hass for the continuing interest she expressed in the project.

I am grateful for the cooperation and valuable assistance given me by the staff of Prindle, Weber and Schmidt. The suggestions of David Chelton have been particularly helpful.

Finally, many, many thanks to my wife Marge for her understanding, patience, and excellent typing of the entire manuscript, and to my son Scot who, despite a total lack of interest in graph theory, gave his mother time to type.

Gary Chartrand
Kalamazoo, Michigan

Contents

Chapter 1
Mathematical Models

Chapter 2
Elementary Concepts of Graph Theory

Chapter 3
Transportation Problems

Chapter 4
Connection Problems

Chapter 5
Party Problems

Chapter 6
Games and Puzzles

Chapter 7
Digraphs and Mathematical Models

Chapter 8
Graphs and Social Psychology

Chapter 9
Planar Graphs and Coloring Problems

*Chapter 10
Graphs and Other Mathematics

Appendix
Sets, Relations, Functions, Proofs

Chapter 1

Mathematical Models

Much of the usefulness and importance of mathematics lies in its ability to treat a variety of situations and problems. The mathematical problems which evolve from the real world have been commonly referred to as "story problems," "word problems," and "application problems." Our goal in this book is to give the real-life problem a mathematical description (or to model it mathematically). Ordinarily, finding a mathematical description is a very complex problem in itself, and there is seldom a unique solution. Indeed, the problem of modeling a real-life situation in a mathematical manner can be so complicated and varied that only the barest introduction is possible at this point.

1.1 Nonmathematical Models

Probably the best way to learn what "mathematical models" are is to look at examples. This is what we shall do in this chapter. We begin, however, by retreating one step to the word "model," since models need not be mathematical. We shall see that the difficulties involved in "constructing" mathematical models may be very similar to the steps in building nonmathematical models.

What exactly does the word "model" mean? Let us consider some uses of this word. Suppose you, the reader, and your husband (perhaps you have a different "model" of a reader!) have received in the mail a brochure which advertises a new land development near your city, including private houses, apartment complexes, and shopping areas. The brochure shows a map of this area. Curved and straight lines represent roads, rectangles represent houses, and other diagrams represent other aspects of this new development. You know, of course, that the map and what it displays is not the actual land development. It is only a *model* of the development.

You have been considering moving from your current apartment, so, with the aid of the map, you and your husband drive to the apartment complex. This drive turns out to be more difficult than anticipated since all the roads leading into the area are dirt roads and very bumpy. (The map didn't mention that!) You arrive at the office of the apartment complex, and in the middle of the room is a large table displaying a miniature *model* of the entire complex. This allows you to see the location of the apartment buildings as well as the office, the swimming pool, the roads, and the children's play area. Several things which are important to you (such as the location of laundry facilities and carports) are not shown in the model, so you ask about these.

You are interested in this new apartment complex and you would like to see what a typical 2-bedroom apartment looks like. So you are directed to a *model* apartment. Although all the apartments available are unfurnished, the model apartment is furnished to help you determine its appearance once you have moved in. However, the model apartment is a bit misleading, for it has been elegantly decorated by a local furniture store while your furniture is perhaps quite ordinary at best.

We have now seen three examples of models. In each case, the model is a representation of something else. Whether the model gives an accurate enough picture of the real entity depends entirely on which features are important to you.

How else is the word "model" used? Perhaps you (a different reader) think of an attractive young woman modeling a swimsuit. In this case, the manufacturer or a department store is trying to sell swimsuits, and rather than displaying them at a counter, they are having a *model* give you an idea of how the suit would look on your wife (or your sister) if you were to buy it. In this case, the model may not give you a very accurate picture of what the swimsuit will look like on the person for whom it is purchased; on the other hand, you may not care.

Another common use of the word "model" is in "model car" or "model airplane." Perhaps you'd like to build a *model* of a 1956 Thunderbird. There are model kits available for this purpose, but these may not be satisfactory if you would like your model to illustrate the dashboard. There must be some limitations on the detail of your model, or otherwise, the only possibility is to purchase your own 1956 Thunderbird.

Intuitively, then, a *model* is something which represents something else. It may be smaller, larger, or approximately the same size as the thing it represents. It illustrates certain key features (but not all features) of the real thing. What features it possesses depends completely on the construction of the model. Ideally, a model should possess certain predetermined characteristics. Whether such a model can be built is often the crucial problem.

Problem Set 1.1

1. Give three examples of models you have encountered. Indicate some pertinent features of each model and describe a feature each model lacks which would be useful for it to possess.

2. Give an example of a model which is (a) larger than, (b) approximately the same size as the thing it represents.

3. Explain the relationship that radio, television, motion pictures, and the theater have with models. What are some of the pertinent differences in how these media model?

4. List some occupations which deal directly with nonmathematical models.

1.2
Mathematical Models

In a mathematical model, we represent or identify a real-life situation or problem with a mathematical system. One of the best-known examples of this representation is plane Euclidean geometry or plane trigonometry, which gives useful results for describing small regions, such as measuring distances. For example, the map of a state would be very useful for determining distances between towns and cities in the state, but in many instances a map of the world would not be as helpful as a globe, say, for calculating distances between certain cities. To indicate how varied mathematical models may be, we present several examples.

Example 1.1

For investment purposes, you have been building apartment houses the past three years. In particular, three years ago you built a 4-apartment building for $100,000. Two years ago, you raised a 6-apartment building for $140,000, and last year you completed an 8-apartment building for $180,000. You are now considering your construction plans for the current year. What kind of mathematical model would represent this situation?

You might observe that in each case the cost C of the apartment building equals the sum of \$20,000 and the product of the number n of apartments and \$20,000, that is,

$$C = 20{,}000 + 20{,}000n$$

for $n = 4, 6, 8$. We might use this formula, then, to model the cost of apartment buildings.

Example 1.2

You left a jug of wine sitting in your car and its temperature is 70°F when you remove it. You place the jug in your refrigerator, where the temperature is 35°. After 30 minutes you observe that the temperature of the wine has dropped to 60°. What mathematical model would represent this situation? Here we might refer to a law of physics which, in this case, states that the rate of change of the temperature T of the wine is proportional to the difference between the temperatures of the wine and of the refrigerator:

$$\frac{dT}{dt} = k(T - 35),$$

where k is the constant of proportionality and t denotes time. In order to arrive at an expression for T, it would be necessary to solve this differential equation.

Example 1.3

You are a member of an organization which has just purchased a new major league baseball franchise: the New Orleans Shrimps. Each of the existing major league teams has agreed to leave unprotected four of their players, and you have the option of purchasing the contract of any of these unprotected players at $75,000 each, provided no more than two players are taken from any one team. You have the records of each of these players. Naturally, you would like to obtain the best players available for your team. What mathematical model would represent this situation? First, you decide to rank pitchers according to their earned-run averages. Then, for non-pitchers, a more complicated formula is adopted. For each player's preceding season, let h be the number of hits, H the number of home runs, r the number of runs scored, R the number of runs batted in, w the number of times the player had the winning hit, f his fielding average, and b his batting average. Then the player's proficiency P is given by

$$P = h + 5H + r + 2R + 10w + 1000f + 1000b.$$

The non-pitchers are then ranked according to their player proficiencies.

Example 1.4

Suppose you are a woman beginning your senior year of college. You realize that you will graduate at the end of the school year, and you have been thinking about your future and what you will do after graduation. You have been considering several possibilities. Although you've heard that the job prospects for new graduates are not excellent, your grades in college have been high, and with an economics major and mathematics minor, you feel that your chances are good for getting a junior executive position with some reputable business firm. This interests you. However, you've been going to college for over three years, and the thought of spending several months traveling after graduation intrigues you. Many of your friends have been talking about this. You have also been thinking about pursuing a graduate degree.

This situation can be represented by the diagram of Figure 1.1, where the "states" S_0, S_1, S_2, and S_3 are as follows:

S_0: You are an undergraduate (initial state).

S_1: You look for a job in the business world.

S_2: You travel with your friends.

S_3: You go to graduate school.

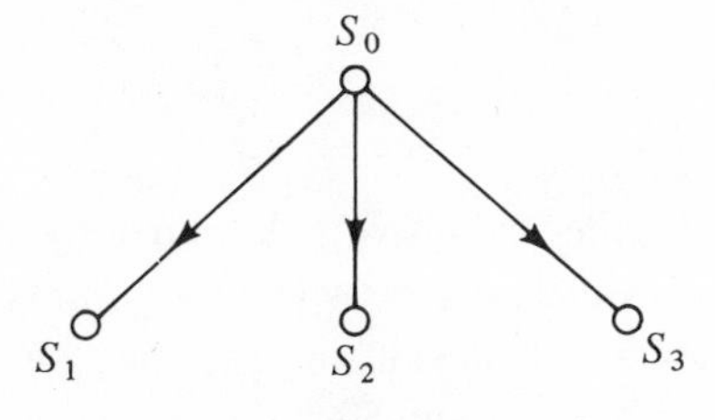

Figure 1.1

You now know what the alternatives are, but what decision should you make? Such a decision requires more information. After thinking this over, you decide the following criteria are most important to you:

How interesting this activity will be.

How this will eventually affect your getting an interesting and challenging job.

How this affects your meeting young men.

How this will affect your immediate financial status.

How much free time this will give you.

You decide to assign 1, 2, or 3 points to each of these factors in each alternative state. The values assigned are:

	S_1	S_2	S_3
interesting activity	2	3	2
eventual job	2	1	3
meeting young men	2	2	3
financial status	3	1	1
free time	1	3	2
	10	10	11

Therefore you decide to go to graduate school. Now where do you apply?—that's another decision.

Example 1.5

You own a rather exclusive golf course in the country, and you are trying to decide what you should charge for a round of golf. You decide to try an experiment. You charge \$11.75 one day and 50 golfers play that day. When you charge \$11.00, a total of 100 people pay to play golf. When you charge \$9.75, the number of golfers totals 150. What mathematical model would represent this situation? First we observe that for $x = 50, 100, 150$, the price p which is being charged is given by

$$p = 12 - 0.0001x^2.$$

Hence, we might choose this formula as our model.

Problem Set 1.2

5. Why is the mathematical model described in Example 1.1 likely to be unsatisfactory?

6. What question would you like the mathematical model of Example 1.2 to answer?

7. You take a quart of milk whose temperature is 40° and you place it on a kitchen counter. The temperature in the kitchen is 75°. After one hour, the temperature of the milk has risen to 45°. What mathematical model would represent this situation?

8. In your opinion, what would be a better mathematical model than the given model for the situation described in Example 1.3? Why is yours preferable?

9. Imagine yourself in the position of the college senior of Example 1.4. List the alternatives you would choose and the factors that are important to you in making a decision. By assigning points as in Example 1.4, make a decision. What do you think of this method of making decisions?

10. Give an example of a decision you are likely to make in the near future. Proceed as in Example 1.4 by listing alternatives, indicating important criteria, and assigning points.

11. What is the most important question a mathematical model for Example 1.5 should answer?

12. The following problem is rather common in a beginning calculus course. A square piece of cardboard, 12 centimeters on a side, is to have squares (all of the same size) cut out of its corners. Then its sides will be folded upward to produce a box with no top. What kind of mathematical model would represent this situation? What question would you like the mathematical model to answer?

13. Suppose you have a fair coin; that is, it is equally likely for heads or tails to appear if the coin is flipped. You are approached by a compulsive gambler who states that if you flip this coin four times and heads and tails appear twice each, then he will pay you $11. On the other hand, if you fail, you will pay him only $10. What kind of mathematical model would you

identify with this situation? What question would you like the mathematical model to answer?

14. You must pay \$10 to play the following game at a carnival. You flip a fair coin three times. When heads comes up the first time, you receive \$5. If heads comes up a second time, you receive an additional \$7. If heads occurs a third time, you receive yet another \$9. Hence, it is possible to receive as much as \$5 + \$7 + \$9 = \$21 (for a net profit of \$11) or as little as nothing (for a net loss of \$10). What kind of mathematical model would you identify with this situation? What question should your model answer for you?

1.3
Graphs

Before proceeding further with illustrations of mathematical models, we pause to introduce the concept of a "graph." As the title of this book indicates, we shall encounter this concept many times.

A *graph* G is a finite nonempty set V together with an irreflexive, symmetric relation R on V. Since R is symmetric, for each ordered pair $(u, v) \in R$, the pair (v, u) also belongs to R. We denote by E the set of symmetric pairs in R. For example, a graph G may be defined by the set

$$V = \{v_1, v_2, v_3, v_4\}$$

together with the relation

$$R = \{(v_1, v_2), (v_1, v_3), (v_2, v_1), (v_2, v_3), (v_3, v_1), (v_3, v_2), (v_3, v_4), (v_4, v_3)\}.$$

In this case,

$$E = \{\{(v_1, v_2), (v_2, v_1)\}, \{(v_1, v_3), (v_3, v_1)\}, \{(v_2, v_3), (v_3, v_2)\}, \{(v_3, v_4), (v_4, v_3)\}\}.$$

In a graph G, we refer to V as the *vertex set*, each element of V being called a *vertex* (the plural is "vertices"). The number of vertices in G is called the *order* of G. Each element of E (that is, each set consisting of two symmetric ordered pairs from R) is called an *edge*, and E itself is called the *edge set* of G. The number of edges in G is called the *size* of G. Hence, $|V|$ = order of G and $|E|$ = size of G.

If G is a graph defined in terms of a vertex set V and a relation R on V, then $(u, v) \in R$ implies $(v, u) \in R$. Hence, $\{(u, v), (v, u)\}$ is an edge of G. It is convenient to denote such an edge by uv (or, equivalently, vu). The edge set E completely determines the relation R; indeed, it is customary to describe a graph in terms of its vertex set and edge set. The graph G illustrated above could then be defined in terms of the sets $V = \{v_1, v_2, v_3, v_4\}$ and $E = \{v_1v_2, v_1v_3, v_2v_3, v_3v_4\}$. Therefore, the order of G is four, as is its size.

Occasionally it is desirable to denote the vertex set and edge set of a graph G by $V(G)$ and $E(G)$, respectively. This is particularly useful when there are two or more graphs under consideration.

Since the empty subset of $V \times V$ is an irreflexive and symmetric relation on V, it follows that the edge set of a graph may be empty, i.e., a graph may have no edges. Of course, by definition, every graph has vertices.

In dealing with graphs, it is often convenient to represent them by means of diagrams. In such a representation, we indicate the vertices by points or small circles, and we represent the edges by line segments or curves joining the two appropriate points. The line segments or curves are drawn so that they pass through no point other than the two points they join. Diagrams of the graph G previously described are given in Figure 1.2. The first diagram uses only straight line segments, while the second diagram employs curved lines. Although the two diagrams look quite different, they contain exactly the same vertices and the same edges, and so they describe the same graph. Notice that in the second diagram, the line segments representing the edges v_1v_2 and v_3v_4 intersect. This is quite permissible (in fact, it may be unavoidable), but you should not confuse this point of intersection with a vertex. As mentioned earlier, for this example, there are four vertices.

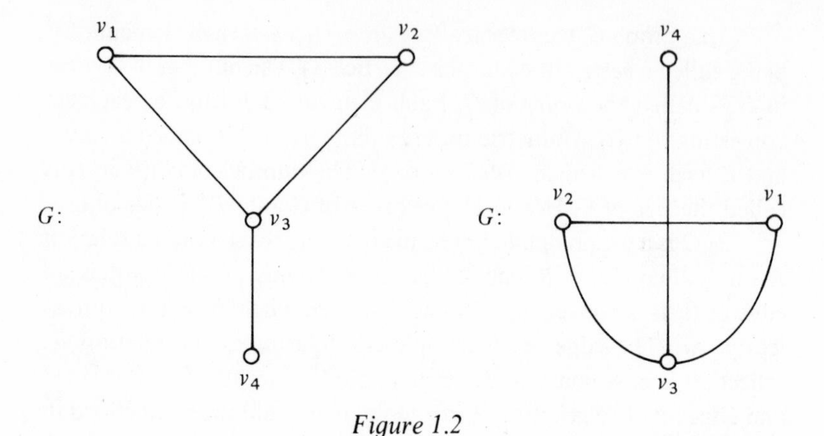

Figure 1.2

Since a diagram of a graph (such as the diagram shown in Figure 1.2) completely describes the graph, it is customary and convenient to refer to the diagram of a graph G as G itself. A few elementary definitions are now in order. (Many of these are inspired by the geometric aspect of graphs.)

If $e = uv \in E(G)$ (i.e., if uv is an edge of a graph G), then we say e *joins* the vertices u and v. Two vertices u and v are *adjacent* in a graph G if $uv \in E(G)$. We say that u and v are *adjacent to* or *adjacent with* each other. If $uv \notin E(G)$, then u and v are *nonadjacent* vertices. If $e = uv \in E(G)$, then u and v are each *incident to* or *incident with e*. If uv and uw are distinct edges of a graph G (i.e., $v \neq w$), then uv and uw are *adjacent* edges. Hence, in the graph G of Figure 1.2, v_1 and v_3 are adjacent, but v_1 and v_4 are not adjacent. The vertex v_3 is incident to the edge v_2v_3, but v_4 is not incident to v_2v_3. The edges v_1v_3 and v_3v_4 are adjacent, but v_1v_2 and v_3v_4 are not adjacent.

Problem Set 1.3

15. Draw the graph with vertex set $V = \{u_1, u_2, u_3, u_4, u_5\}$ and edge set $E = \{u_1u_2, u_1u_4, u_1u_5, u_2u_3, u_3u_5, u_4u_5\}$. What is the corresponding irreflexive symmetric relation R on V?

16. Draw a graph G with vertex set $V = \{u_1, u_2, u_3, u_4, u_5\}$ and edge set E such that $|E|$ is as large as possible. Determine E.
17. If a graph G has order 3, what are the possible sizes for G?
18. What is the maximum possible size of a graph of (a) order 3; (b) order 4; (c) order 5; (d) order n, where n is a positive integer?
19. Does an example exist of a graph of order 3 such that every two vertices are adjacent and every two edges are adjacent? Does such a graph of order 4 exist?
20. Is there a graph G of order five or more such that every vertex of G is incident with at least one edge, but no two edges are adjacent?
21. Let $n \geq 2$ be an integer. If G is a graph of order n, what is the minimum size possible for G (in terms of n) if G contains a vertex which is adjacent to all other vertices of G?
22. Give an example of a graph G of positive size with the property that every vertex is incident with every edge.
23. Give an example of a graph exhibiting the properties that:
    (a) every vertex is adjacent to two vertices; and
    (b) every edge is adjacent with two edges.
24. If V is a nonempty set, why does it follow that the empty subset of $V \times V$ is an irreflexive, symmetric relation on V? Is the relation also transitive?

1.4
Graphs as Mathematical Models

The construction of mathematical models may take many forms and may involve many areas of mathematics. One area of mathematics particularly well-suited to model building is graph theory. In this section we present examples of situations and describe the appropriate graphs that serve as mathematical models. At this point, we make no attempt to consider detailed problems. We shall delay discussion of problems until Chapter 3.

Example 1.6

A grade-school teacher wishes to make a seating chart for her class. How she constructs the seating chart may depend on which students are friends of each other. The class can be "pictured" by means of a graph, where the vertices represent the students and friendship between two students is indicated by an edge between the appropriate vertices.

Example 1.7

Several army platoons are deployed in various locations in preparation for a battle. Communication is handled by battery-powered telephone. Two platoons can communicate directly with each other if they are sufficiently close. A model of this situation would be a graph where the vertices represent the platoons and direct communication between two platoons is represented by an edge between the two appropriate vertices.

Example 1.8

A number of islands are located in the Pacific Ocean off the California coast. Suppose a line of ferryboats operates from the mainland to certain of the islands. Suppose further that boats travel between several islands as well. This situation can be represented by means of a graph, where the vertices denote the islands and the mainland (one vertex for each island and one for the mainland). Two vertices are joined by an edge if you can travel by boat directly between the land areas.

Problem Set 1.4

25. In the graph of Example 1.6, what observation could you expect to make concerning the vertex associated with the "most popular child" in the class? What would you expect to find concerning a vertex associated with a new child in the class? Is it possible that a graph constructed at the beginning of the school year might be different from one constructed at the end of the school year? How could this happen?

*26. Suppose we associate a graph G with a college class in the following manner. The vertices of G correspond to the students in the class, while two vertices of G are adjacent if and only if they correspond to two students having the same major. Can you describe the appearance of G?

27. In Example 1.7, it might be important for two platoons to communicate (indirectly, if not directly). How is it possible to determine by looking at the graph of Example 1.7 whether every two platoons can communicate with each other?

28. What would be an important question to ask concerning the situation described in Example 1.8? Could this question be answered with the aid of the corresponding graph?

29. The Student Council in a certain high school consists of 15 members. Ten different committees in school are made up of Student Council members. Some committees may have only a few members, while others may have many. Some Student Council members may belong to no committees, and other

members may belong to several. Give examples of two graphs which describe this situation. Can you think of conditions which may make one graph more useful than the other?

*30. We can generalize Exercise 29 in the following way. Let U be a finite nonempty set, and let $S_1, S_2, \ldots, S_n$ be a collection of nonempty subsets of U. Find two examples of graphs which describe this situation.

31. Give an example of a real-life situation which can be represented by a graph. Draw the graph as it may appear.

1.5
Directed Graphs as Mathematical Models

A *directed graph* D (often called a *digraph*) is a finite nonempty set V together with an irreflexive relation R on V. As with graphs, the elements of V are called *vertices*. Each ordered pair in R is referred to as a *directed edge* or *arc* (the word "edge" is not used in digraphs). For consistency with the notation introduced for graphs, we shall denote the relation by E rather than by R when dealing with digraphs.

Since the defining relation of a digraph D need not be symmetric, it follows that if (u, v) is an arc of D, then (v, u) need not be an arc of D. This situation would be indicated in a diagram of D by drawing a line segment or curve between the points representing u and v and inserting an arrowhead that "directs" the line segment from u to v. Should both (u, v) and (v, u) be arcs of D, then, ordinarily, we would draw two curves (which do not cross) between u and v, and place an arrowhead on each curve in opposite directions.

If we let $V_1 = \{v_1, v_2, v_3, v_4\}$ and $E_1 = \{(v_1, v_2), (v_2, v_3), (v_3, v_2)\}$, then we have described a digraph, say D_1. The digraph D_1 can be indicated pictorially, as in Figure 1.3. Again we shall follow the custom of referring to a diagram of a digraph as the digraph itself.

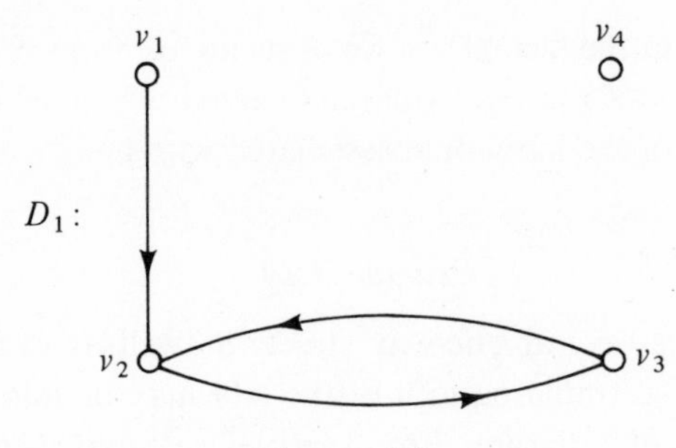

Figure 1.3

It may happen that the relation defining a digraph D is symmetric. We refer to such digraphs as *symmetric digraphs.* Of course, symmetric digraphs are then graphs. The only real difference between a symmetric digraph and a graph is how they are represented pictorially. For example, Figure 1.4 shows a symmetric digraph D and its graphical counterpart G.

There are certain situations for which digraphs yield a more acceptable mathematical model than graphs can provide. We consider two examples.

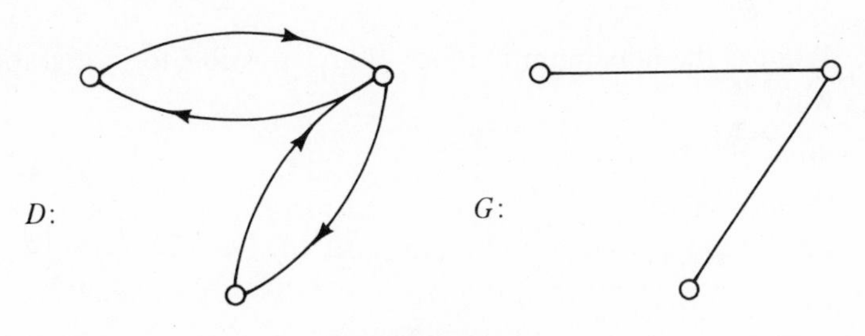

Figure 1.4

Example 1.9

A large New York business firm has a rather complex structure. We can represent this structure by a digraph D. Namely, we identify a vertex of D with each individual working in the firm. Then we draw an arc from vertex u to vertex v if the individual associated with v is a subordinate of the individual associated with u.

Example 1.10

A city has several one-way streets as well as two-way streets. The traffic pattern of the city may be indicated by means of a digraph. For example, we could represent the street intersections by vertices, and introduce an arc from u to v if it is possible to drive legally from the intersection associated with u to the intersection associated with v without passing through any other intersection.

Problem Set 1.5

32. Draw the digraph whose vertex set is $V = \{v_1, v_2, v_3, v_4, v_5, v_6\}$ and whose arc set is $E = \{(v_1, v_3), (v_2, v_3), (v_3, v_4), (v_4, v_1), (v_4, v_3), (v_5, v_6)\}$.

33. What is the maximum number of arcs possible for a digraph with

    (a) 3 vertices?

    (b) 4 vertices?

    (c) 5 vertices?

    (d) n vertices?

34. What property would you expect the relation associated with the digraph of Example 1.9 to have? Describe a simpler digraph than the one given which could model the situation of Example 1.9 equally well.

35. With regard to Example 1.10, what important property would you expect of the traffic pattern? Could this property be determined from the digraph?

36. Give an example of a situation which could be represented better by a digraph than by a graph.

1.6
Networks as Mathematical Models

Just as there are instances when digraphs are more suitable than graphs as mathematical models for certain situations, there are occasions when neither graphs nor digraphs are entirely appropriate as mathematical models, although graphs or digraphs appear to be involved. In this section we consider some other alternatives.

By a *network* we mean a graph or digraph together with a function which maps the edge set into the set of real numbers. (The word "network" is used because of its connection with electrical networks.) A network resulting from a graph is called an *undirected network*, while a network resulting from a digraph is called a *directed network*. Examples of each are shown in Figure 1.5.

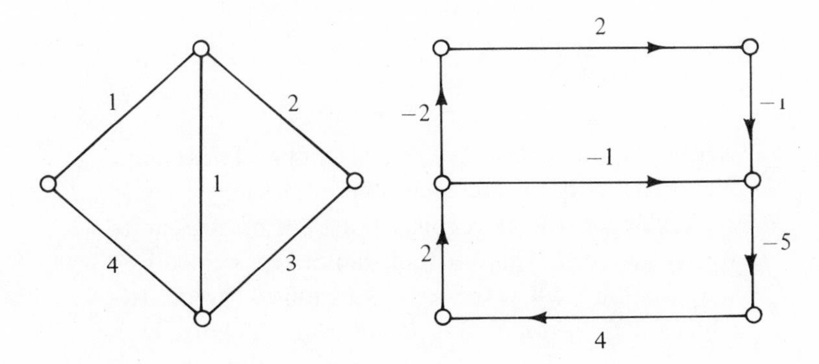

Figure 1.5

A *signed graph* S is an undirected network whose functional values are $+1$ or -1. Since a positive or negative sign is attached to every edge of S, it is natural to refer to each edge of a signed graph as a *positive edge* or *negative edge*. For example, if

$$V = \{v_1, v_2, v_3\},$$
$$E = \{v_1v_2, v_1v_3, v_2v_3\},$$

and

$$f = \{(v_1v_2, +1), (v_1v_3, -1), (v_2v_3, -1)\},$$

then the resulting signed graph can be represented in one of two ways, as shown in Figure 1.6.

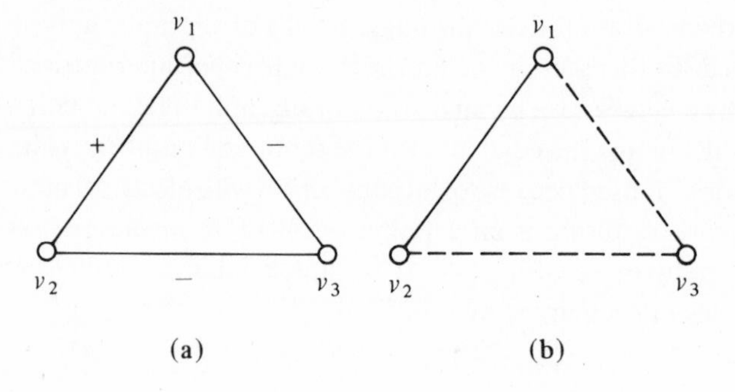

Figure 1.6

Example 1.11

A neighborhood consists of several families. Two families may be friendly toward each other, unfriendly, or may be indifferent toward (or may not even be acquainted with) each other. This situation can be represented by a signed graph S, where the vertices are joined by a positive edge if the corresponding families are friendly, by a negative edge if the corresponding families are unfriendly, and by no edge otherwise.

Undirected networks whose functional values are positive integers often serve as mathematical models. There are two common ways to represent such undirected networks. For example, if

$$V = \{v_1, v_2, v_3\},$$
$$E = \{v_1v_2, v_1v_3, v_2v_3\},$$

and

$$f = \{(v_1v_2, 2), (v_1v_3, 1), (v_2v_3, 3)\},$$

then the resulting undirected network can be represented as shown in Figure 1.7(a) or Figure 1.7(b).

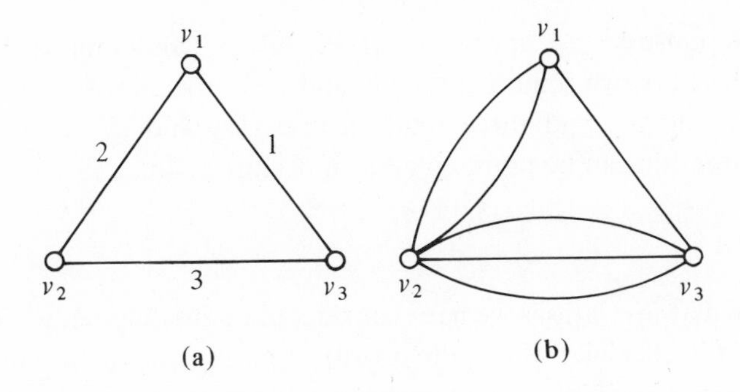

Figure 1.7

If such an undirected network is represented as a set of points in the plane and the points are joined by an integral number of curves or line segments [as in Figure 1.7(b)], then the network is called a *multigraph*. Let M be a multigraph with edge set E and associated function f. If $uv \in E$ and $f(uv) = n$, where n is a positive integer, then we say u and v are joined by n edges, and we refer to these edges as *multiple edges*.

Example 1.12

Let v_1, v_2, and v_3 be three villages, and suppose a road runs between every two villages. The quality of the roads and the distances between villages make the walking times between the villages as follows:

between v_1 and v_2, two days;

between v_1 and v_3, one day;

between v_2 and v_3, three days.

This situation can be represented as in Figure 1.7(a).

Example 1.13

Suppose v_1, v_2, and v_3 are three villages, and suppose there are two roads between v_1 and v_2, one road between v_1 and v_3, and three roads between v_2 and v_3. This situation can be represented as in Figure 1.7(b).

In all the relations we have considered in this chapter, we have assumed irreflexivity. It is quite possible that in the situation under discussion, the relation is not irreflexive. In this case, we refer to the ordered pair (u, u) as a *loop.* If we remove the restriction "irreflexive" in the definition of graph, we call the result a *loop-graph. Loop-digraphs, loop-networks,* and *loop-multigraphs* are defined analogously. Loop-multigraphs are also called *pseudographs.*

Let

$$V = \{v_1, v_2, v_3, v_4\}$$

and

$$E = \{(v_1, v_2), (v_2, v_3), (v_3, v_2), (v_3, v_3), (v_4, v_4)\}.$$

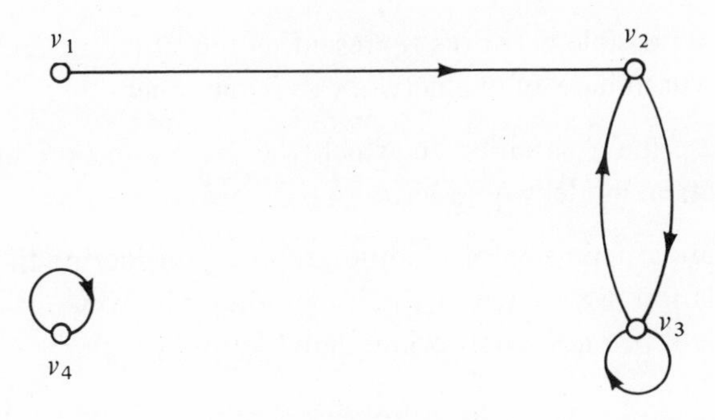

Figure 1.8

This is a loop-digraph, which could be drawn as indicated in Figure 1.8.

Problem Set 1.6

37. Give an example (of a type different from Example 1.11) in which a signed graph is an appropriate mathematical model.

38. Why is the situation described in Example 1.13 more conducive to representation by a multigraph than is the situation described in Example 1.12?

39. In Example 1.10 we indicated how a digraph can help describe the traffic pattern of a city having one-way and two-way streets. Explain how a network may be more useful than a digraph in describing this situation for a letter-carrier.

40. A businessman is driving from Michigan City, Indiana, to O'Hare Airport in Chicago. He reaches an intersection where he has the option of continuing along the main highway, where the traffic increases significantly, or taking a different route which is faster but requires paying a toll of $1.40. Give

two possible networks representing the situation, and explain the usefulness of one network over the other.

41. Describe a situation in which the most appropriate mathematical model would have loops.

42. State a definition of "loop-digraph." Considering the type of relation associated with a loop-digraph, what other term might be more appropriate than "loop-digraph"?

Suggestions for Further Reading

Textbooks providing an introduction to mathematical models often contain a treatment of finite mathematics. One good case in point is the book by Malkevitch and Meyer [4], which also includes a discussion of graphs. Indeed, the book by Kemeny, Snell, and Thompson [2], who first brought finite mathematics to the limelight, includes topics on mathematical models.

There are two texts on graph theory, Ore [5] and Wilson [7], which incorporate several examples of graphs used as mathematical models. The first chapter of Harary [1] contains several examples as well. Advanced treatments of mathematical models of various types are given by Maki and Thompson [3] and by Roberts [6]; however, these require a more extensive background in a variety of mathematical areas.

[1] F. Harary, *Graph Theory*. Addison-Wesley, Reading, Mass. (1969).

[2] J. G. Kemeny, J. L. Snell, and G. L. Thompson, *Introduction to Finite Mathematics*, third edition. Prentice-Hall, Englewood Cliffs, N.J. (1974).

[3] D. P. Maki and M. Thompson, *Mathematical Models and Applications.* Prentice-Hall, Englewood Cliffs, N.J. (1973).

[4] J. Malkevitch and W. Meyer, *Graphs, Models, and Finite Mathematics.* Prentice-Hall, Englewood Cliffs, N.J. (1974).

[5] O. Ore, *Graphs and Their Uses.* Random House, New York, N.Y. (1963).

[6] F. S. Roberts, *Discrete Mathematical Models.* Prentice-Hall, Englewood Cliffs, N.J. (1976).

[7] R. J. Wilson, *Introduction to Graph Theory.* Academic Press, New York, N.Y. (1972).

Chapter 2

Elementary Concepts of Graph Theory

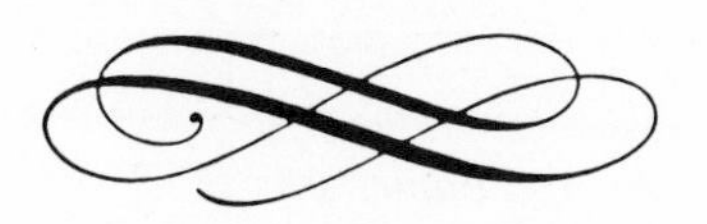

To this point, we have introduced several general situations for which graphs or related concepts serve as mathematical models. As we proceed, we shall ask questions pertaining to particular situations and their resultant models. In order to deal with these models in some detail, we must become more familiar with the terminology of graph theory and some of the basic results. We shall investigate these topics in this chapter.

2.1
The Degree of a Vertex

We have already introduced two numbers associated with a graph G, namely the order and the size. Now we define a collection of numbers associated with G. Let v be a vertex of G. The number of edges of G incident with v is called the *degree* of v in G. The degree of v is denoted by $\deg_G v$, or simply $\deg v$ if the graph is clear by context. For the graph G of Figure 2.1, $\deg v_1 = 1$, $\deg v_2 = 2$, $\deg v_3 = 3$, $\deg v_4 = 2$, and $\deg v_5 = 0$.

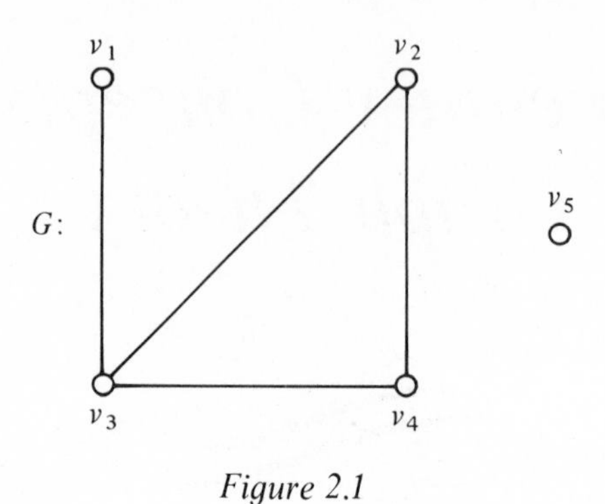

Figure 2.1

By a (p, q) *graph* we mean a graph having order p and size q. The graph G of Figure 2.1 is a $(5, 4)$ graph. We might observe that for this graph, the sum of the degrees of its vertices is 8, which, in this case, equals $2q$. This is no mere coincidence, as we now show.

Theorem 2.1

For any graph G, the sum of the degrees of the vertices of G equals twice the number of edges of G. Symbolically, if G is a (p, q) graph with vertices $v_1, v_2, \ldots, v_p$, then

$$\sum_{i=1}^{p} \deg v_i = 2q.$$

Proof

When summing the degrees of the vertices of a graph G, we count each edge of G twice, once for each of the two vertices incident with the edge. ∎

A vertex is called *even* or *odd* according to whether its degree is even or odd. The graph G of Figure 2.1 has two odd vertices and three even vertices. The following result is a consequence (or corollary) of Theorem 2.1.

Theorem 2.2

Every graph contains an even number of odd vertices.

Proof

Let G be a graph. If G contains no odd vertices, then the result follows immediately. Suppose that G contains k odd vertices; denote them by $v_1, v_2, \ldots, v_k$. If G contains even vertices as well, then denote these by $u_1, u_2, \ldots, u_n$. By Theorem 2.1,

$$(\deg v_1 + \deg v_2 + \cdots + \deg v_k) + (\deg u_1 + \deg u_2 + \cdots + \deg u_n) = 2q,$$

where q is the number of edges in G. Since each of the numbers $\deg u_1, \deg u_2, \ldots, \deg u_n$ is even, $(\deg u_1 + \deg u_2 + \cdots + \deg u_n)$ is even, so we have

$$(\deg v_1 + \deg v_2 + \cdots + \deg v_k) = 2q - (\deg u_1 + \deg u_2 + \cdots + \deg u_n) \quad \text{is even.}$$

However, each of the numbers $\deg v_1, \deg v_2, \ldots, \deg v_k$ is odd. Therefore, k must be even; that is, G has an even number of odd vertices. If G has no even vertices, then we have $(\deg v_1 + \deg v_2 + \cdots + \deg v_k) = 2q$, from which we again conclude that k is even. ▮

If every vertex of a graph G has the same degree r, we say that G is *regular of degree r* or is *r-regular*. A graph is *complete* if every two of its vertices are adjacent. A complete graph of order p is $(p - 1)$-regular and is denoted by K_p. Five complete graphs are shown in Figure 2.2.

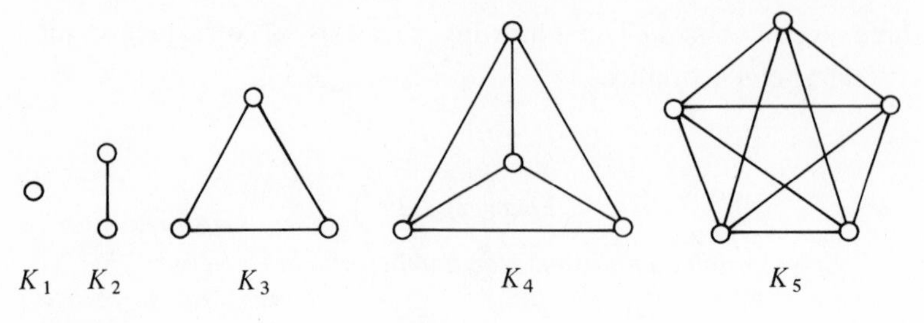

Figure 2.2

Problem Set 2.1

1. For the graph G of Figure 2.3, determine deg v_i for $i = 1, 2, \ldots, 8$ and determine its order p and size q. Then, illustrate Theorem 2.1 by verifying that $\sum_{i=1}^{8} \deg v_i = 2q$.

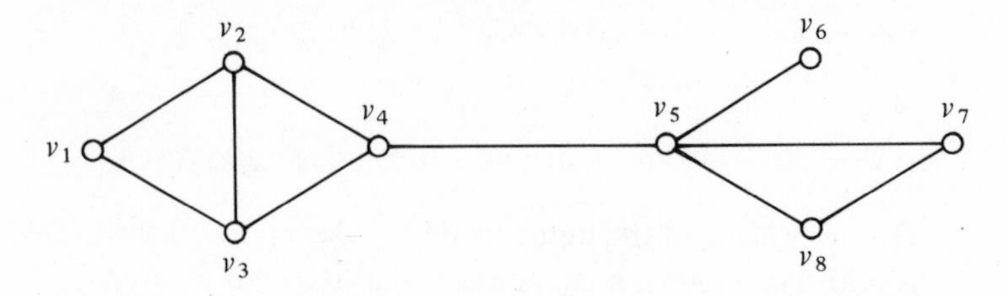

Figure 2.3

2. Show that a graph G cannot exist with vertices of degrees 2, 3, 3, 4, 4, and 5.

3. Show that a graph G cannot exist with vertices of degrees 2, 3, 4, 4, and 5.

*4. Show that a graph G cannot exist with vertices of degrees 1, 3, 3, and 3.

5. Suppose we know the degrees of the vertices of a graph G. Is it possible to determine the order and size of G? Explain.

*6. Suppose we know the order and size of a graph G. Is it possible to determine the degrees of the vertices of G? Explain

7. Give an example of a graph
   (a) having no odd vertices.
   (b) having no even vertices.
   (c) having exactly one odd vertex.
   (d) having exactly one even vertex.
   (e) having exactly two odd vertices.
   (f) having exactly two even vertices.

8. Let p and n be nonnegative integers such that n is even and $0 \leq n < p$. Give an example of a graph G of order p containing exactly n odd vertices. Is it necessary to stipulate that n is even?

9. If G is a (p, q) r-regular graph, express q in terms of p and r. What is q if $G = K_p$?

10. Give an example of a
    (a) 0-regular graph which is not complete.
    (b) 1-regular graph which is not complete.
    (c) 2-regular graph which is not complete.
    (d) 3-regular graph which is not complete.

11. Let m and n be nonnegative integers such that $m \neq n$. Find an example of a graph G such that each vertex has degree m or n. (*Note*: Your example must be general. Do not assign specific values to m and n.)

*12. A graph G of order $p\,(\geq 2)$ is called *perfect* if no two of its vertices have equal degrees. Prove that "no graph is perfect." If we replace "(≥ 2)" by "(≥ 1)," does the conclusion change?

*13. Suppose you and your husband attended a party with three other married couples. Several handshakes took place. No one shook hands with himself (or herself) or with his (or her) spouse, and no one shook hands with the same person more than once. After all the handshaking was completed, suppose you asked each person, including your husband, how many hands he or she had shaken. Each person gave a different answer.

(a) How many hands did you shake?

(b) How many hands did your husband shake?

2.2
Isomorphic Graphs

In every area of mathematics, it is important to know whether two objects under investigation are the same (in some sense) or are different. For example, the numbers 2 and 6/3 are considered to be the same, or equal, but they are certainly not identical. We now wish to determine what conditions must hold for two graphs to be "equal." The importance of knowing this equality lies in the fact that if G_1 and G_2 are two equal graphs which are models of two situations, then there is something basically similar about the two situations.

Intuitively, two graphs G_1 and G_2 are the same if it is possible to redraw one of them, say G_2, so it appears identical to G_1. For example, the graphs G_1 and G_2 of Figure 2.4 have this property.

We refer to two equal graphs as "isomorphic graphs." We now give a more formal definition of this concept. Let G_1 and G_2 be two graphs. By an *isomorphism* from G_1 to G_2 we mean a one-to-one mapping $\phi : V(G_1) \rightarrow V(G_2)$ from $V(G_1)$ onto $V(G_2)$ such that two vertices u_1 and v_1 are adjacent in G_1 if and only if the vertices $\phi(u_1)$ and $\phi(v_1)$ are adjacent in G_2. We then say that G_1 and G_2 are

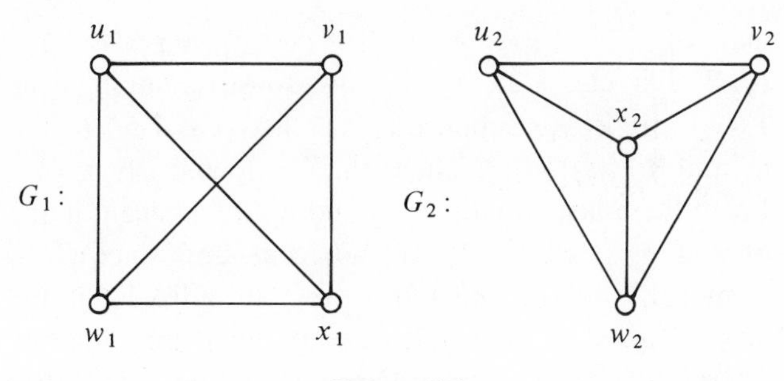

Figure 2.4

isomorphic if an isomorphism exists from G_1 to G_2. If ϕ is an isomorphism from G_1 to G_2, then the inverse mapping ϕ^{-1} (see Exercise A.25, page 255) from $V(G_2)$ to $V(G_1)$ also satisfies the definition of isomorphism. If G_1 and G_2 are isomorphic graphs, we can say that G_1 is isomorphic to G_2 and that G_2 is isomorphic to G_1.

An important property of isomorphism is contained in the following theorem.

Theorem 2.3

The relation "is isomorphic to" is an equivalence relation on the set of all graphs.

Proof

The fact that the relation "is isomorphic to" is reflexive follows immediately. We need only observe that if G is a graph and the mapping $\phi: V(G) \to V(G)$ is defined by $\phi(v) = v$ for all $v \in V(G)$, then ϕ is an isomorphism from G to G, i.e., G is isomorphic to G.

Suppose G_1 is isomorphic to G_2; that is, suppose ϕ is an isomorphism from G_1 to G_2. Define the inverse mapping $\phi^{-1}: V(G_2) \to V(G_1)$ by $\phi^{-1}(v_2) = v_1$ if $\phi(v_1) = v_2$. By Exercise A.25, ϕ^{-1} is a one-to-one mapping from $V(G_2)$ onto $V(G_1)$. Suppose $u_2, v_2 \in V(G_2)$, and $\phi^{-1}(u_2) = u_1$ and $\phi^{-1}(v_2) = v_1$. Then $\phi(u_1) = u_2$ and $\phi(v_1) = v_2$. From these last equalities, u_2 and v_2 are adjacent if and only if $\phi(u_1)$ and $\phi(v_1)$ are adjacent, and since G_1 is isomorphic to G_2, $\phi(u_1)$ and $\phi(v_1)$ are adjacent if and only if $u_1 = \phi^{-1}(u_2)$ and $v_1 = \phi^{-1}(v_2)$ are adjacent. Therefore, u_2 and v_2 are adjacent if and only if $\phi^{-1}(u_2)$ and $\phi^{-1}(v_2)$ are adjacent. This shows that G_2 is isomorphic to G_1, i.e. "is isomorphic to" is a symmetric relation.

We still must show that the relation is transitive. Suppose that G_1 is isomorphic to G_2 and that G_2 is isomorphic to G_3. Hence there exist isomorphisms $\alpha: V(G_1) \to V(G_2)$ and $\beta: V(G_2) \to V(G_3)$. Consider the composite mapping $\beta \circ \alpha$. By Theorems A.4 and A.5 (page 254), $\beta \circ \alpha$ is a one-to-one mapping from $V(G_1)$ onto $V(G_3)$. Let $u_1, v_1 \in V(G_1)$. Suppose that $\alpha(u_1) = u_2$ and $\alpha(v_1) = v_2$, and that $\beta(u_2) = u_3$ and $\beta(v_2) = v_3$. Since α and β are isomorphisms, u_1 and v_1 are adjacent if and only if $\alpha(u_1) = u_2$ and $\alpha(v_1) = v_2$ are adjacent; and u_2 and v_2 are adjacent if and only if $\beta(u_2) = u_3$ and $\beta(v_2) = v_3$ are adjacent. Thus, u_1 and v_1 are adjacent if and only if $u_3 = (\beta \circ \alpha)(u_1)$ and $v_3 = (\beta \circ \alpha)(v_1)$ are adjacent. This completes the proof that $\beta \circ \alpha$ is an isomorphism. Hence, G_1 is isomorphic to G_3. ∎

By Theorem A.2 (page 248), it follows that the equivalence relation "is isomorphic to" partitions the set of all graphs into equivalence classes. Hence, two graphs which belong to the same equivalence class are isomorphic, while two graphs belonging to different equivalence classes are not isomorphic (that is, they are considered different graphs).

If G_1 and G_2 are isomorphic graphs, then, by definition, there exists a one-to-one mapping ϕ from $V(G_1)$ onto $V(G_2)$. This implies that $V(G_1)$ and $V(G_2)$ have the same number of elements; that is, G_1 and G_2 have the same order. Let u_1 and v_1 be two vertices of G_1, and suppose that $\phi(u_1) = u_2$ and $\phi(v_1) = v_2$. Then u_1 and v_1 are adjacent in G_1 if and only if u_2 and v_2 are adjacent in G_2, or in other words, u_1v_1 is an edge of G_1 if and only if u_2v_2 is an edge of G_2. This implies that G_1 and G_2 have the same size. However, if two graphs have the same order and the same size, it does not necessarily follow that the graphs are isomorphic. For example, the two graphs of Figure 2.5 have order six and size nine, but they are not isomorphic.

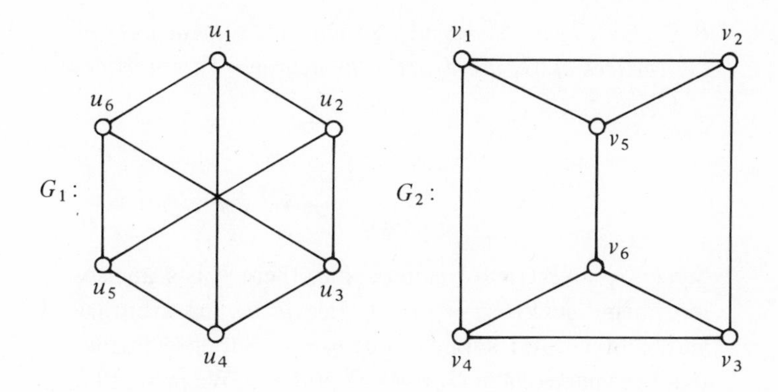

Figure 2.5

It may seem a difficult problem to show that the graphs G_1 and G_2 of Figure 2.5 are not isomorphic, for evidently, we must verify that *every* one-to-one mapping from $V(G_1)$ onto $V(G_2)$ [or from $V(G_2)$ to $V(G_1)$] fails to be an isomorphism. However, we can simplify the problem immensely by making some pertinent observations. In the case of the graphs G_1 and G_2 of Figure 2.5, consider any one-to-one mapping ϕ from $V(G_1)$ onto $V(G_2)$. The vertices v_1, v_2, and v_5 of G_2 are pairwise adjacent, and ϕ must map three vertices of G_1 into v_1, v_2, and v_5. If ϕ is an isomorphism, then two vertices of G_1 are adjacent if and only if the two image vertices of G_2 under ϕ are adjacent. This implies that the three vertices of G_1 whose images are v_1, v_2, and v_5 also must be pairwise adjacent; however, G_1 does not contain three pairwise adjacent vertices. Hence there is no isomorphism from $V(G_1)$ to $V(G_2)$ and G_1 is not isomorphic to G_2.

A very useful *necessary* condition for a graph G_1 to be isomorphic to a graph G_2 is presented next.

Theorem 2.4

If G_1 and G_2 are isomorphic graphs, then the degrees of the vertices of G_1 are exactly the degrees of the vertices of G_2.

Proof

Since G_1 and G_2 are isomorphic, there exists an isomorphism $\phi: V(G_1) \to V(G_2)$. Let u be an arbitrary vertex of G, and suppose $\deg u = n$. Suppose further that the image of u in G_2 is v, i.e., $\phi(u) = v$. We prove that $\deg v = n$.

Since $\deg u = n$, the graph G_1 contains vertices $u_1, u_2, \ldots, u_n$ which are adjacent to u, while every other vertex of G_1 is not adjacent to u. Let $\phi(u_i) = v_i$ for $i = 1, 2, \ldots, n$. Then v is adjacent to each of the vertices $v_1, v_2, \ldots, v_n$, since ϕ is an isomorphism. Furthermore, these are the only vertices adjacent to v, since u is adjacent to x in G_1 if and only if v is adjacent to $\phi(x)$ in G_2. Thus, $\deg v = n$.

Because a vertex of G_1 and its image vertex of G_2 have the same degree, this establishes the theorem. ▌

We again emphasize that Theorem 2.4 gives a *necessary* condition for two graphs to be isomorphic—*not a sufficient* condition. That is, the vertices of two graphs may have exactly the same degrees, but may not be isomorphic. (For example, G_1 and G_2 of Figure 2.5 are not isomorphic.) On the other hand, if the degrees of the vertices of a graph G_1 and the degrees of the vertices of a graph G_2 are not the same, then by Theorem 2.4, G_1 and G_2 are *not* isomorphic.

It follows that there is only one graph of order one (necessarily having size zero); that is, if G_1 and G_2 are both graphs of order one, then they are isomorphic. Similarly, there is only one graph of order two and size zero, and only one graph of order two and size one. However, there are three graphs of order four and size three, shown in Figure 2.6. No two graphs in Figure 2.6 are isomorphic, but any other graph of order four and size three is isomorphic to one of the graphs of Figure 2.6. We can state this in another way: Among the graphs of order four and size three, there are three isomorphism classes. Thus, if we have four or more graphs of order four and size three, two or more of these graphs must belong to the same class.

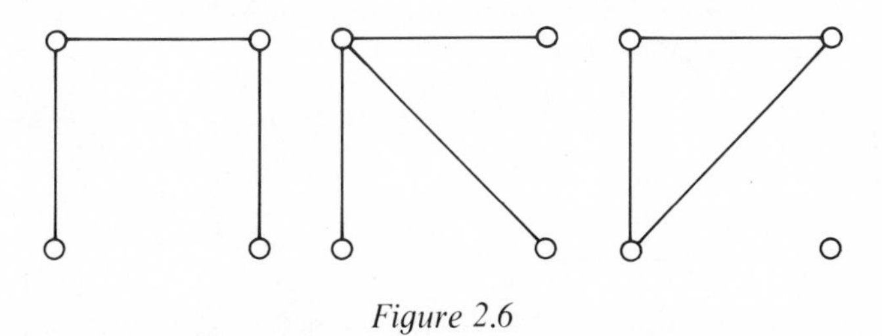

Figure 2.6

The preceding discussion illustrates the following celebrated result from the field of combinatorics. (*Note*: For a real number x, the number $\{x\}$ denotes the smallest integer greater than or equal to x.)

The Pigeonhole Principle

Let S be a finite set consisting of n elements, and let $S_1, S_2, \ldots, S_k$ be a partition of S into k subsets. Then at least one subset S_i, $1 \le i \le n$, contains at least $\{n/k\}$ elements.

Hence, if there are three equivalence classes of (4, 3) graphs (that is, $k = 3$), and we have four (4, 3) graphs (that is, $n = 4$), then there must be at least $\{4/3\} = 2$ graphs in the same equivalence class.

Problem Set 2.2

14. Show that the graphs G_1 and G_2 of Figure 2.4 are isomorphic by redrawing G_2 in such a way that it appears exactly like G_1.

15. (a) Show that the graphs G_1 and G_2 of Figure 2.4 are isomorphic by showing that an isomorphism ϕ exists from G_1 to G_2.

    (b) Show that the graphs G_1 and G_2 of Figure 2.4 are isomorphic by showing that an isomorphism θ exists from G_2 to G_1.

16. Show that the graphs F_1 and F_2 of Figure 2.7 are isomorphic

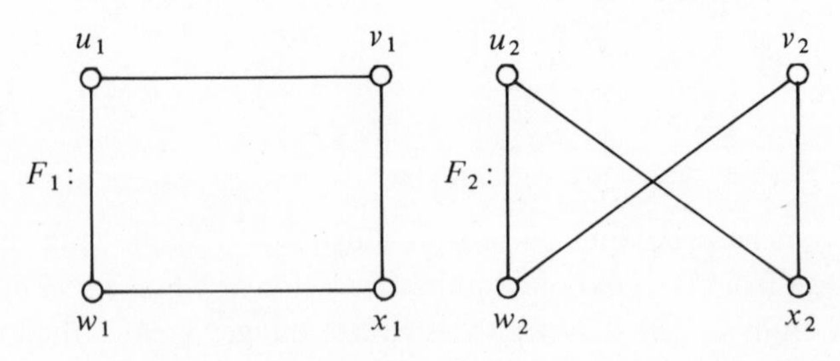

Figure 2.7

by proving the existence of an isomorphism from F_1 to F_2 or of an isomorphism from F_2 to F_1.

17. Let G_1 be a graph whose vertices have degrees 2, 2, 3, 3, 4, and 4, and let G_2 be a graph whose vertices have degrees 2, 3, 3, 3, 3, and 4. Can G_1 and G_2 be isomorphic? Explain.

18. Show that no two graphs of Figure 2.6 are isomorphic.

*19. Show that the (8, 8) graphs F_1 and F_2 of Figure 2.8 are not isomorphic.

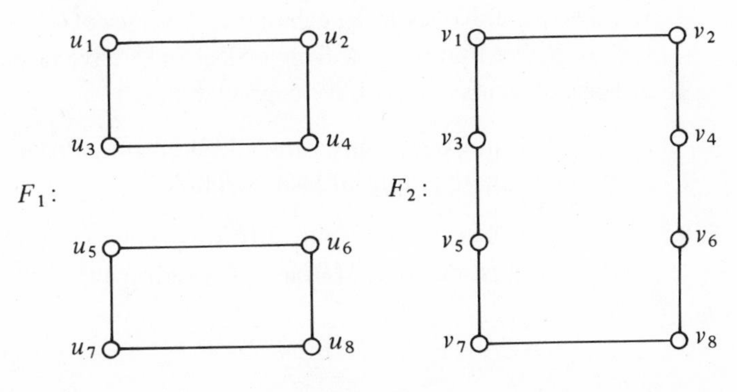

Figure 2.8

20. The degrees of the vertices of the non-isomorphic graphs F_1 and F_2 of Figure 2.8 are 2, 2, 2, 2, 2, 2, 2, and 2. Does an (8, 8) graph F_3 exist with that same property, such that F_3 is isomorphic to neither F_1 nor F_2? Explain.

21. Show that the graph G_3 of Figure 2.9 (page 40) is isomorphic to exactly one of the graphs of Figure 2.5.

22. Give an example of two non-isomorphic (4, 2) graphs. Verify that these graphs are not isomorphic.

*23. Prove the Pigeonhole Principle. (*Hint*: Try a proof by contradiction.)

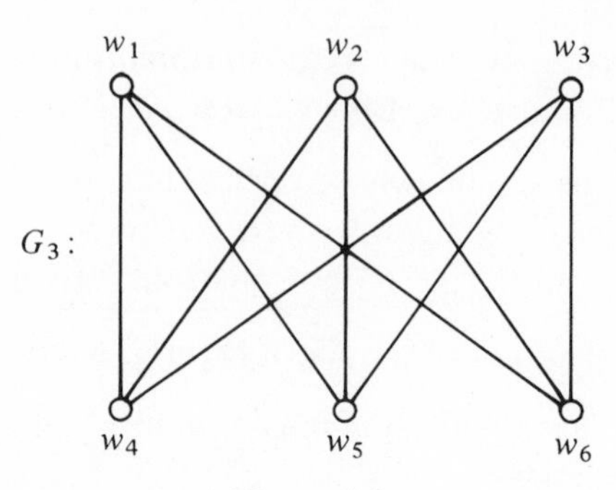

Figure 2.9

*24. Let G be a graph of order 9 such that each vertex of G has degree 5 or 6. Prove that at least five vertices of G have degree 6 or at least six vertices of G have degree 5.

*25. Let G_1, G_2, and G_3 be any three (4, 2) graphs. Prove that at least two of these graphs are isomorphic.

*26. Which pairs of graphs in Figure 2.10 are isomorphic and which pairs are not isomorphic? (*Hint*: Try redrawing them.)

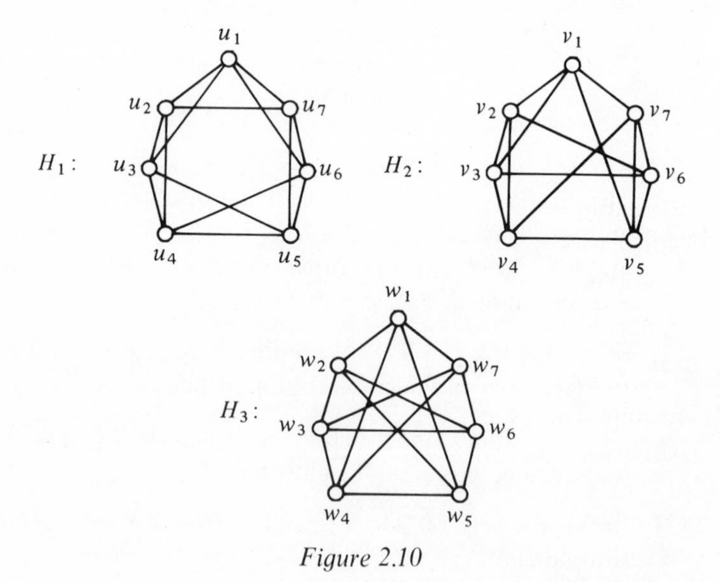

Figure 2.10

2.3
Connected Graphs

Probably the most important class of graphs is the class of connected graphs. In this section we discuss connected graphs together with some related concepts.

Let G be a graph. A graph H is a *subgraph* of G if $V(H) \subseteq V(G)$ and $E(H) \subseteq E(G)$. If a graph F is isomorphic to a subgraph H of G, then F is also called a subgraph of G. Figure 2.11 shows a graph G and a subgraph H.

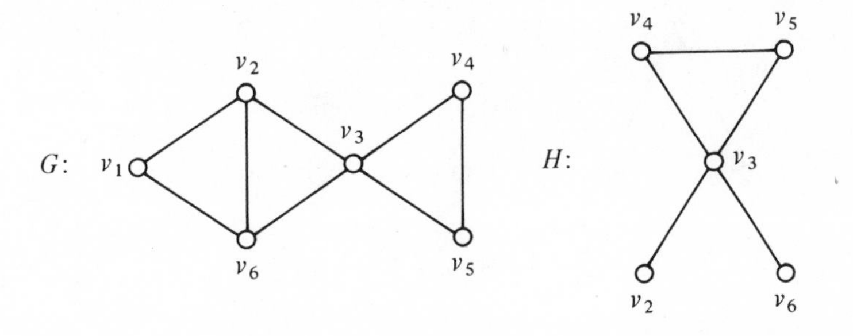

Figure 2.11

Let u and v be vertices of a graph G. A *u-v walk* in G is an alternating sequence of vertices and edges of G, beginning with u and ending with v, such that every edge joins the vertices immediately preceding it and following it. For example, v_3, v_3v_2, v_2, v_2v_6, v_6, v_6v_3, v_3, v_3v_4, v_4, v_4v_5, v_5, v_5v_4, v_4 is a v_3-v_4 walk in the graph G of Figure 2.11. Observe that the edge v_4v_5 appears twice in this walk. We need only list the vertices in a walk, for the edges are then obvious. The walk just described can therefore be expressed more simply as $v_3, v_2, v_6, v_3, v_4, v_5, v_4$.

A *u-v trail* in a graph is a u-v walk which does not repeat any edge. The v_3-v_4 walk described above is not a v_3-v_4 trail; however, v_3, v_2, v_6, v_3, v_4 is a v_3-v_4 trail in the graph G of Figure 2.11. A *u-v path* is a u-v walk (or u-v trail) which does not repeat any vertex. Again, in the graph G of Figure 2.11, v_3, v_5, v_4 is a v_3-v_4 path.

Two vertices u and v in a graph G are *connected* if $u = v$, or if $u \neq v$ and a u-v path exists in G. A graph G is *connected* if every two vertices of G are connected; otherwise, G is *disconnected.*

A connected subgraph H of a graph G is called a *component* of G if H is not contained in any connected subgraph of G having more vertices or edges than H. For example, the graph of Figure 2.12 has four components. If a graph has only one component, the graph is connected.

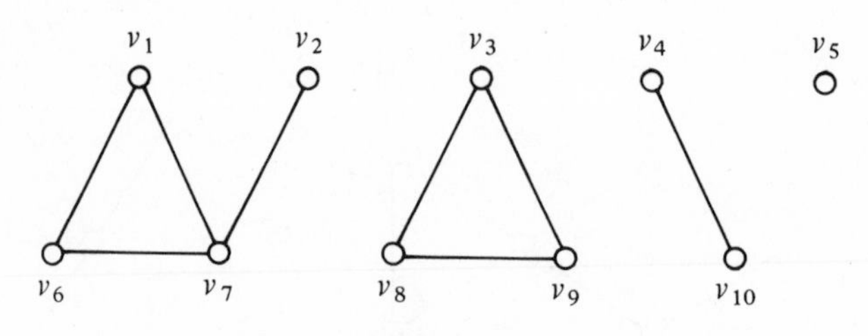

Figure 2.12

A u-v trail in which $u = v$ and which contains at least three edges is called a *circuit*. In other words, a circuit must end at the same vertex with which it began. A circuit which does not repeat any vertices (except the first and last) is called a *cycle*. For example, in the graph G of Figure 2.13, $v_1, v_2, v_3, v_5, v_2, v_6, v_1$, is a circuit but is not a cycle, while v_2, v_4, v_3, v_5, v_2 is a cycle (as well as a circuit).

By definition, a trail is an alternating sequence of vertices and edges, although we have agreed to represent a trail by a sequence of vertices. The sets of vertices and of edges determined by a trail produce a subgraph; it is also customary to refer to this subgraph as a trail. For example, $v_1, v_2, v_5, v_3, v_2, v_6$ is a trail in the graph G of Figure 2.13. If we define the subgraph H of G by $V(H) = \{v_1, v_2, v_5, v_3, v_6\}$ and $E(H) = \{v_1v_2, v_2v_5, v_5v_3, v_3v_2, v_2v_6\}$, then

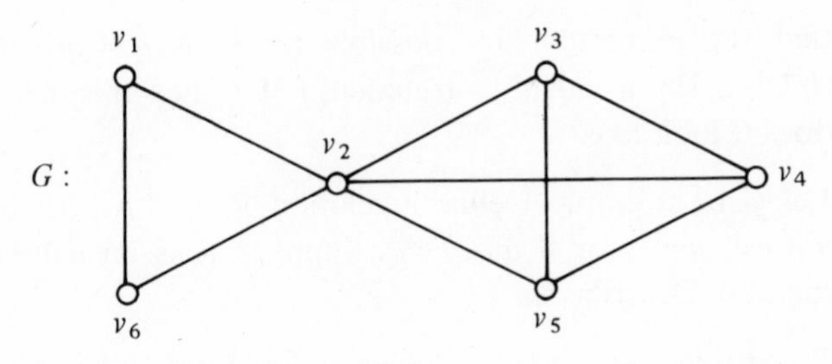

Figure 2.13

H is also called a trail in G. More generally, it is customary to consider the subgraph consisting of the vertices and edges of a trail, path, circuit, or cycle as the respective trail, path, circuit, or cycle.

Problem Set 2.3

27. Give an example of a disconnected graph with four components where each component is complete.

28. Give an example of a disconnected graph with three components where no two components are isomorphic.

29. Give an example of a disconnected graph with three components where every two components are isomorphic.

30. Let n and p be integers such that $1 \leq n \leq p$. Provide an example of a graph of order p having n components.

31. Is it possible for a graph to have more components than vertices? Explain.

32. Let G be a graph of order 13 containing three components. Show that at least one component of G has at least five vertices.

*33. Let G be a graph of even order p (i.e., $p = 2n$ for some positive integer n) such that G has two complete components. Prove that the minimum size possible for G is $q = (p^2 - 2p)/4$. (*Hint*: Try a calculus argument.) If G has this size, what does G look like?

*34. Let G be a graph. Define a relation R on $V(G)$ as follows: $u\,R\,v$ if $u = v$ or if $uv \in E(G)$. Suppose R is an equivalence relation. Describe G.

*35. Prove that a graph G is connected if and only if for every two vertices u and v of G, there exists a u-v walk in G.

*36. Let G_1 and G_2 be isomorphic graphs. Prove that if G_1 is connected, then G_2 is connected.

*37. Let G be a graph of order $p\,(\geq 2)$, and suppose that for every vertex v of G, $\deg v \geq (p-1)/2$. Prove that G is connected. (*Hint*: Try a proof by contradiction. To begin, assume that $\deg v \geq (p-1)/2$ for every vertex v of G and that G is disconnected. Since G is disconnected, G has two or more components. What can be said about the number of vertices in each component?)

*38. Let G be a graph of order $p\,(\geq 2)$, and suppose that for every vertex v of G, $\deg v \geq (p-2)/2$. Show that G need not be connected if p is even (see Exercise 37).

39. In the graph G of Figure 2.14, give an example of a circuit C which is not a cycle. Describe the subgraph H of G whose vertices and edges belong to C.

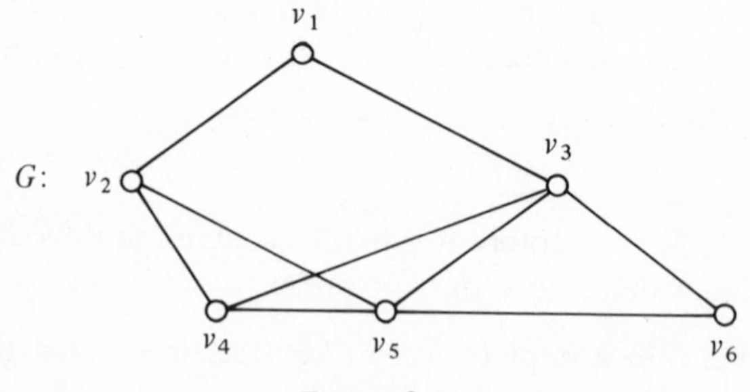

Figure 2.14

40. In the graph G of Figure 2.14, give an example of
    (a) a trail which is not a path;
    (b) a path;
    (c) a cycle.

*41. Prove that every trail contains a path.

*42. Prove that every circuit contains a cycle.

43. Consider the sequence of vertices $v_1, v_2, v_3, v_4, v_5, v_2, v_1$ of the graph G of Figure 2.13. Is this a circuit, a cycle, or neither? Explain.

*44. Let G be a graph, and let R denote the relation "is connected to" on the set $V(G)$. Show that R is an equivalence relation. Determine the equivalence classes.

2.4
Cut-Vertices and Bridges

We now introduce a class of vertices and a class of edges which are important and similar in many ways.

If e is an edge of a graph G, then $G - e$ is the subgraph of G possessing the same vertex set as G and having all edges of G except e. If v is a vertex of a graph G containing at least two vertices, then $G - v$ is the subgraph of G whose vertex set consists of all vertices of G except v and whose edge set consists of all edges of G except those incident with v. Figure 2.15 illustrates these concepts.

A vertex v in a connected graph G is called a *cut-vertex* if $G - v$ is disconnected. The vertex v_3 of the graph G in Figure 2.15 is a cut-vertex; however, no other vertex of that graph is a cut-vertex.

Now we consider the related concept for edges. An edge e in a connected graph G is called a *bridge* if $G - e$ is disconnected. The

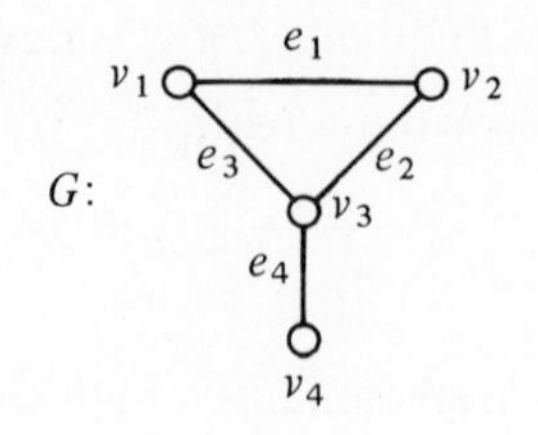

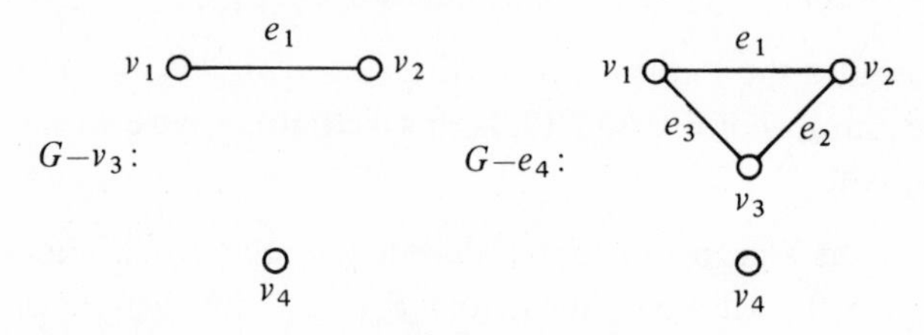

Figure 2.15

edge e_4 of the graph G in Figure 2.15 is a bridge, but no other edge of that graph is a bridge.

If v is a cut-vertex of a connected graph G, then $G - v$ contains two or more components. However, if e is a bridge of G, then $G - e$ has exactly two components. The following theorem shows which edges of a graph are bridges.

Theorem 2.5

Let G be a connected graph. An edge e of G is a bridge of G if and only if e does not lie on any cycle of G.

Proof

Let e be a bridge of G. We prove the desired result by a contradiction argument. Suppose $e = uv$ and e does lie on a cycle, say $C: u, v, w, \ldots, x, u$ (that is, w follows v on C and x precedes u). The graph $G - e$ contains a u-v

path, namely $u, x, \ldots, w, v$, so that u is connected to v. We now show that $G - e$ is connected.

Let u_1 and v_1 be any two vertices of $G - e$; we show that $G - e$ contains a u_1-v_1 path. Since G is connected, there is a u_1-v_1 path P in G. If the edge e does not lie on P, then P is also a path in $G - e$ and u_1 is connected to v_1 in $G - e$. Suppose the edge e lies on P. Then the path P can be expressed as $u_1, \ldots, u, v, \ldots, v_1$, or $u_1, \ldots, v, u, \ldots, v_1$. In the first case, u_1 is connected to u and v is connected to v_1 in $G - e$, and in the second case, u_1 is connected to v and u is connected to v_1. We have already observed that u is connected to v in $G - e$. By Exercise 2.44, the relation "is connected to" is an equivalence relation on $V(G - e)$; hence the relation is transitive, implying that u_1 is connected to v_1. Therefore, if e belongs to a cycle, then $G - e$ is connected and e is not a bridge. This produces the desired contradiction.

Conversely, suppose $e = uv$ is an edge which lies on no cycle of G. Again, we give a proof by contradiction. Assume that e is not a bridge. Then $G - e$ is connected, and thus, there is a u-v path P in $G - e$. However, P together with e produces a cycle in G containing e. Hence, we have a contradiction. ▮

Problem Set 2.4

45. In the graph G of Figure 2.16 (page 48), determine the cut-vertices and the bridges of G.

46. Give an example of a connected graph G containing a vertex v such that $G - v$ has four components.

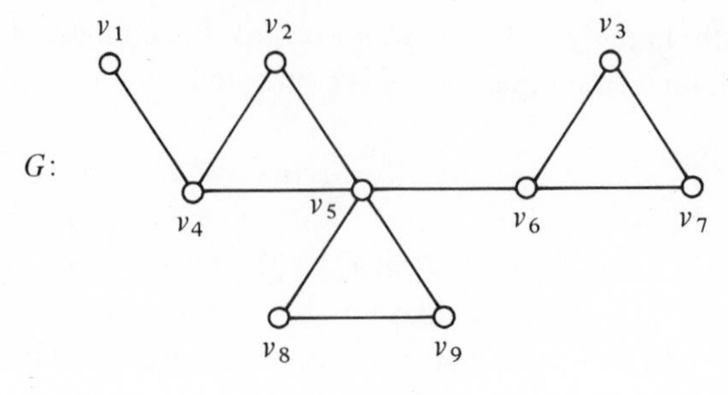

Figure 2.16

47. Let G be a graph of order 11 and let e be a bridge of G and v be a vertex of G.
    (a) Show that there exists a component of $G - e$ containing at least six vertices.
    (b) Show that there need not exist a component of $G - v$ containing at least six vertices.

*48. Let G be a connected graph containing only even vertices. Prove that G cannot contain a bridge. (*Note*: In any negative-type result, a good proof technique to try is proof by contradiction.)

49. Let G be a connected graph containing only even vertices. Show that it is possible for G to contain cut-vertices.

50. By Theorem 2.5, if e is an edge of a connected graph G such that e does not lie on any cycle of G, then e must be a bridge of G. Show that if v is a vertex of G such that v does not lie on any cycle of G, then v need not be a cut-vertex of G.

51. Is Theorem 2.5 still true if the word "cycle" is replaced by "circuit"? Verify your answer.

52. Restate Theorem 2.5 using the phrase "necessary and sufficient."

53. Give an example of a connected, 3-regular graph containing a bridge.

54. Give an example of a connected graph containing more bridges than cut-vertices.

55. Give an example of a connected graph containing more cut-vertices than bridges.

56. One possible definition of a cut-vertex of a disconnected graph is the following: A vertex v of a disconnected graph G is a cut-vertex if v is a cut-vertex of a component of G. Can you give another, yet equivalent, definition? Give two (equivalent) definitions of a bridge in a disconnected graph.

57. Give an example of a graph of order 5 such that every edge is a bridge.

*58. Let G be a connected graph, and let u, v, and w be three vertices of G. Suppose that every u-w path contains v. What property does v have? Prove your assertion.

*59. If G is a connected graph not isomorphic to K_2, and if e is a bridge of G, show that e is incident with a cut-vertex in G.

*60. Determine whether the following statement is true: If G is a connected graph with a cut-vertex, then G has a bridge. If true, give a proof; if false, give an example of a connected graph with a cut-vertex and no bridges.

Suggestions for Further Reading

In this chapter we have taken a brief look at some of the basic concepts of graph theory. We shall encounter several more as we continue through the text; however, we shall not investigate any particular topic in great depth. There are several readable texts that present introductory accounts of the mathematical theory of graphs. Three such examples are books by Wilson [3], Harary [2], and

Behzad, Chartrand and Lesniak-Foster [1]. The first of these is the most elementary of the three; the latter two discuss a larger variety of graph-theoretic concepts and problems.

[1] M. Behzad, G. Chartrand, and L. Lesniak-Foster, *Graphs & Digraphs*. Prindle, Weber & Schmidt, Boston. To be published.

[2] F. Harary, *Graph Theory*. Addison-Wesley, Reading, Mass. (1969).

[3] R. J. Wilson, *Introduction to Graph Theory*. Academic Press, New York (1972).

Chapter 3

Transportation Problems

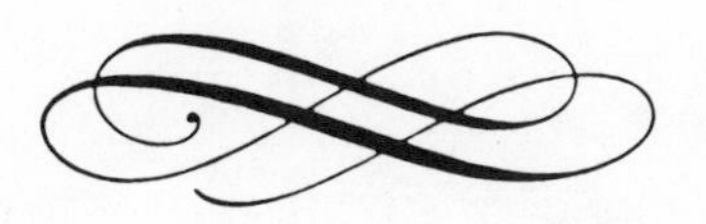

We have introduced and discussed some of the basic terminology related to graphs, and we have seen a few examples of how graphs may serve as mathematical models representing certain situations. In this chapter we begin our investigation of problems which can be solved with the aid of graphs used as mathematical models. The first collection of problems we consider belongs to the general area of transportation problems.

3.1
The Königsberg Bridge Problem: An Introduction to Eulerian Graphs

Probably the earliest example of a problem making use of graphs (or related concepts) as mathematical models occurred in 1736. This is the famous Königsberg Bridge Problem. We quote from Newman [6]:

The Königsberg Bridge Problem

IN THE TOWN of Königsberg there were in the 18th century seven bridges which crossed the river Pregel. They connected two islands in the river with each other and with opposite banks. The townsfolk had long amused themselves with this problem: Is it possible to cross the seven bridges in a continuous walk without recrossing any of them?

Figure 3.1 shows a schematic diagram of Königsberg, with the land areas denoted by A, B, C, and D.

The situation in Königsberg can be conveniently represented by a multigraph (see p. 21), as shown in Figure 3.2. The vertex set corresponds to the land areas and each two vertices are joined by a number of edges equal to the number of bridges joining the corresponding land areas.

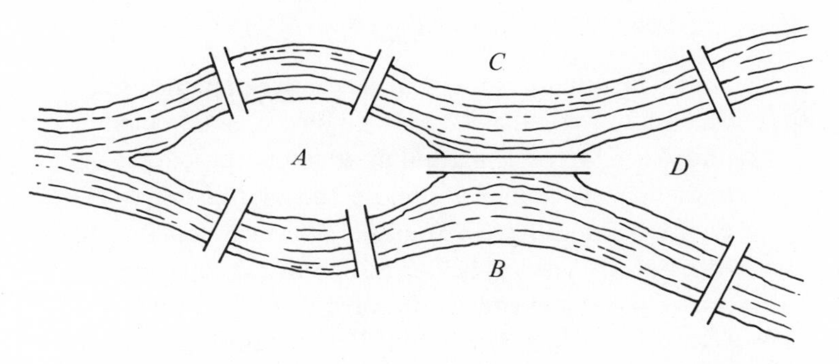

Figure 3.1

The Königsberg Bridge Problem is essentially the problem of determining whether the multigraph M of Figure 3.2 has a trail (possibly a circuit) containing all its edges. You might use a trial-and-error method, and you would probably reach the conclusion that no such trail exists. However, how do you *prove* that no such trail exists? We present a proof of this fact.

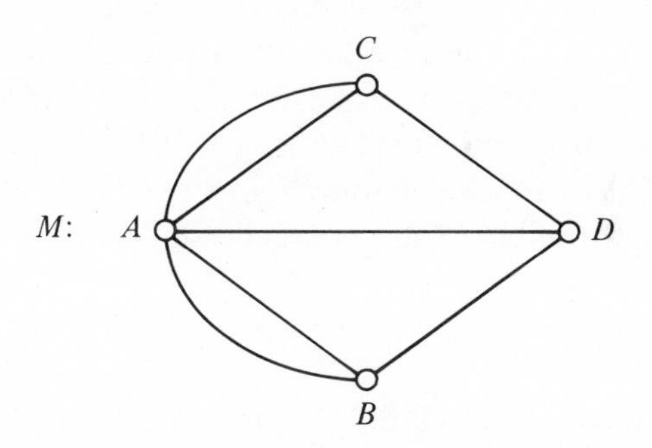

Figure 3.2

Theorem 3.1
Solution of the Königsberg Bridge Problem

The multigraph M of Figure 3.2 has no trail containing all edges of M.

Proof

Notice first that this theorem is negative in nature, since we wish to show that M does *not* have a certain kind of trail. Therefore, it is natural to attempt a proof by contradiction, as follows. Suppose the multigraph M of Figure 3.2 does have a trail, say P, which contains all edges of M. Then P begins at one of the four vertices A, B, C, or D and ends at one of A, B, C, or D (the same vertex from which P started if P is a circuit). Now there are at least two vertices among A, B, C, and D such that P neither begins nor ends at that vertex. Hence there is at least one vertex among B, C, and D at which P neither begins nor ends. Let us denote such a vertex by v.

Notice that each of the vertices B, C, and D has degree 3. Thus, after some edge on P enters vertex v for the first time and some other edge on P leaves vertex v, there is exactly one edge incident with v which does not yet belong to P. Now v must be entered along trail P once again via the edge incident with v which was not yet used. However, upon arriving at v the second time, we find no edges remaining to exit, so that P must terminate at v. This is impossible since P does not end at v. Hence, no such trail P exists, producing the desired contradiction. ∎

As a note to the preceding theorem and proof, we repeat the statement that any theorem claiming the nonexistence of some quantity, or, more generally, having some negative-sounding aspect to it, is commonly proved by a contradiction argument.

The Königsberg Bridge Problem was initially solved by the famous Swiss mathematician Leonhard Euler (1707–1783). The type of trail sought in the Königsberg Bridge Problem has given rise, in a very natural way, to a class of graphs (actually multigraphs) bearing the name of Euler.

A circuit containing all the vertices and edges of a multigraph M is called an *eulerian circuit* in M. A graph containing an eulerian circuit is called an *eulerian graph,* while a multigraph containing an eulerian circuit is an *eulerian multigraph.* The graph in Figure 3.3 is eulerian; C: $u_1, u_2, u_3, u_4, u_5, u_3, u_6, u_7, u_1, u_3, u_7, u_2, u_6, u_1$ is one eulerian circuit.

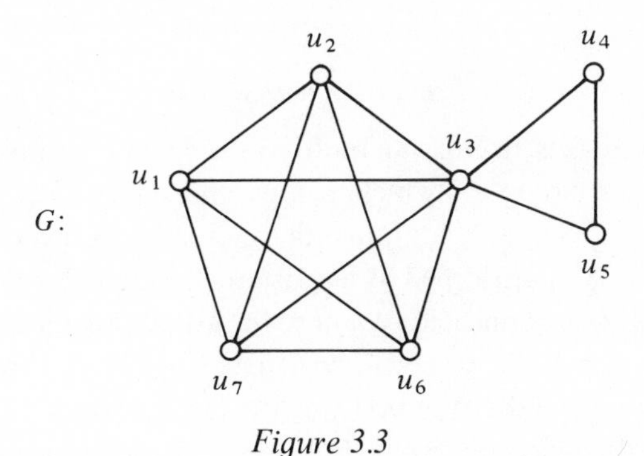

Figure 3.3

The following theorem provides a very simple solution to the problem of determining which graphs and multigraphs are eulerian.

Theorem 3.2

A multigraph G is eulerian if and only if G is connected and every vertex of G is even.

The proof of Theorem 3.2, although not extremely difficult, is somewhat lengthy. Thus, before giving the proof, we illustrate the procedure used on the eulerian graph G of Figure 3.3. Consider the vertex u_1, say. We begin a trail P at u_1 and continue the trail as long as possible. If we are fortunate, P will be an eulerian circuit; however, it may happen that we obtain the circuit P: $u_1, u_2, u_3, u_6, u_7, u_1, u_3, u_7, u_2, u_6, u_1$. In this case, P is not an eulerian circuit, since it does not contain all edges and vertices of G. However, u_3 is a vertex of P that is incident with edges not on P. We now begin a trail P_1 at u_3 which contains edges not belonging to P. If we continue P_1 as long as we can, one possible choice of P_1 would be u_3, u_4, u_5, u_3. We now insert P_1 into P at the first place where u_3 is encountered, obtaining $u_1, u_2, u_3, u_4, u_5, u_3, u_6, u_7, u_1, u_3, u_7, u_2, u_6, u_1$, which is an eulerian circuit.

Proof of Theorem 3.2

Suppose G is an eulerian multigraph. Then G contains an eulerian circuit C, which begins and ends, say, at the vertex v. Since C contains all vertices of G, every two vertices of G are joined by a trail (and therefore by a path), so that G is connected. We now show that every vertex of G is even. First we consider a vertex u different from v. Since u is neither the first nor the last vertex of C, each time u is encountered it is entered via some edge and exited via another edge; hence, each occurrence of u in C increases the degree of u by two. Thus, u has even degree. In the case of the vertex v, each occurrence of v except the first and the last contributes two to its degree, while the initial and final occurrences of v in C contribute one each to the degree of v. Therefore, every vertex of G has even degree.

We now consider the converse statement. Assume that G is a connected multigraph and every vertex in G is even. We show that G is eulerian. Select a vertex v of G, and begin a trail P at v. We continue this trail as long as possible until we reach a vertex w such that the only edges incident with w already belong to P; hence, P cannot be continued, and we must stop. We claim that $w = v$. To establish this claim, suppose that $w \neq v$. On each occasion that w is encountered prior to the last time, we use one edge to enter w and another edge to exit from w. When w is encountered for the final time on P, only one edge is used—namely, to enter w. Hence, an odd number of edges incident with w appears on P. However, since w has even degree, there must be at least one edge incident with w that does not belong to P. This implies that P can be continued and therefore cannot terminate at w, if $w \neq v$. We conclude that $w = v$, and P is actually a circuit. If the circuit P contains all the edges of G, then P is an eulerian circuit of G and G is an eulerian multigraph.

Suppose the circuit P does not contain all the edges of G. Since G is connected, there must be at least one vertex u on P that is incident with edges not on P. Remove the edges of P from G and consider the resulting multigraph H. Since P does not contain all the edges of G, the multigraph H has edges. Furthermore, every vertex belonging to P is incident with an even number of edges of P; hence, every vertex in H has even degree. Let H_1 be the component of H containing the vertex u. If we begin a trail P_1 in H_1 at u and continue this trail as long as possible, then, as before, P_1 must end at u (that is, P_1 must be a circuit). Now it is possible to form a circuit C_1 of G, beginning and ending at v, which has more edges than P. We do this by taking the circuit P and inserting the circuit P_1 at a place where u occurs.

If C_1 contains all the edges of G, then C_1 is an eulerian circuit and G is an eulerian multigraph. If C_1 does not contain all the edges of G, then we may continue the above procedure until we finally obtain an eulerian circuit of G. ▮

We now consider an analogous concept. If a multigraph G has a trail, not a circuit, containing all vertices and edges of G, then

G is called a *traversable graph* and the trail is called an *eulerian trail.* Figure 3.4 shows a traversable graph and $P: v_1, v_2, v_4, v_3, v_2, v_5, v_4$ is an eulerian trail.

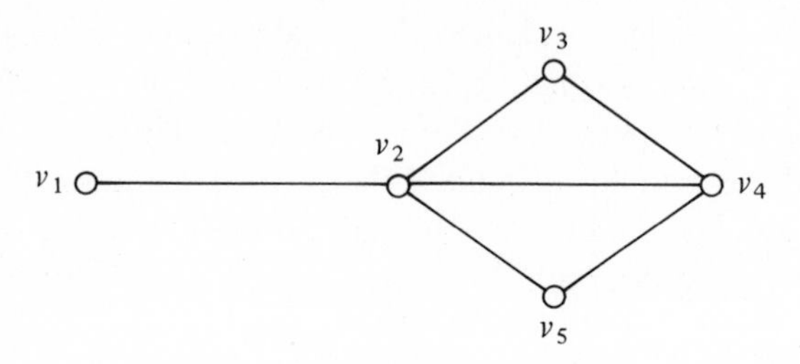

Figure 3.4

The following theorem indicates precisely which graphs are traversable.

Theorem 3.3

A multigraph G is traversable if and only if G is connected and has exactly two odd vertices. Furthermore, any eulerian trail of G begins at one of the odd vertices and ends at the other odd vertex.

We can now see that the multigraph M of Figure 3.2 is neither eulerian nor traversable, so there is no trail in M containing all edges of M. This fact gives us another solution of the Königsberg Bridge Problem.

An interesting property of eulerian and traversable multigraphs is that once the vertices have been drawn, we can draw the edges in one "continuous" motion. In other words, the edges of a connected multigraph can be drawn "without lifting the pencil from the paper" provided the number of odd vertices is zero or two.

Eulerian and traversable graphs and multigraphs find numerous applications in solving mazes, puzzles, and similar problems. We give two such examples now.

Example 3.1

Figure 3.5 shows the floor plan of a house with various doorways leading between rooms, and between several rooms and the outside. Is it possible to start someplace (either in a room or outdoors) and walk through every doorway once and only once?

We use a multigraph as a mathematical model of this situation. We first associate a vertex with each room and a vertex with the outdoors. Every two vertices are joined by a number of edges equal to the number of doorways between the corresponding rooms (or the room and outside). The answer to the question now depends on

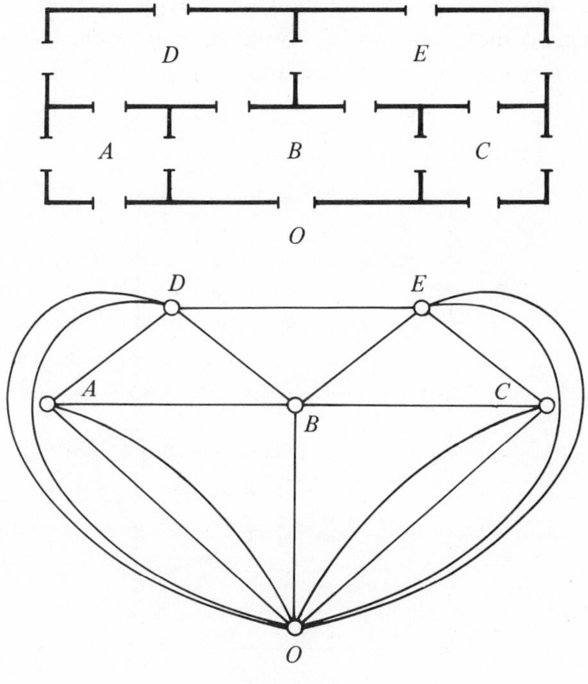

Figure 3.5

whether the multigraph is eulerian, traversable, or neither. We see, however, that vertices *B*, *D*, *E*, and *O* are odd, so the multigraph of Figure 3.5 is neither eulerian nor traversable. Hence, it is not possible to walk through every doorway once and only once.

Example 3.2

Suppose you hold a summer job as a highway inspector. Among your responsibilities, you must periodically drive along the several highways shown schematically in Figure 3.6 and inspect the roads for debris and possible repairs. If you live in town *A*, is it possible to find a round trip, beginning and ending at *A*, which takes you over each section of highway exactly once? If you were to move to town *B*, would it be possible to find such a round trip beginning and ending at *B*?

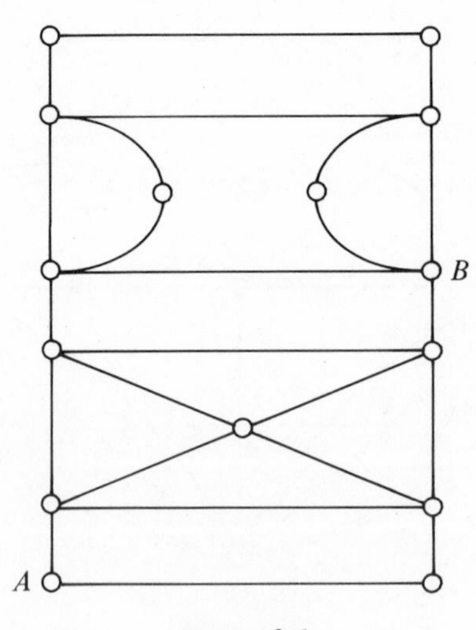

Figure 3.6

To solve this problem, we need only regard the highways as a graph. Observe that the graph of Figure 3.6 is connected and every vertex is even. Therefore, the graph is eulerian and contains an eulerian circuit C. The circuit C contains every edge of the graph exactly once, so a round trip must exist containing each section of highway exactly once. Since a circuit may begin at any vertex of the circuit, there are round trips beginning at either A or B. (A round trip beginning at B, however, will pass through B at some intermediate time before coming to an end.)

Problem Set 3.1

1. Classify the graphs in Figure 3.7 as eulerian, traversable, or neither.

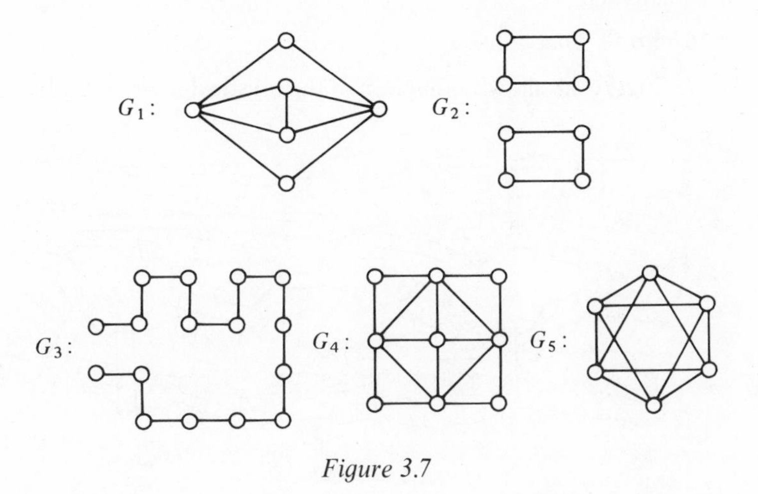

Figure 3.7

2. Give an example of a graph of order 10 which is
   (a) eulerian.
   (b) traversable.
   (c) neither eulerian nor traversable.

3. Let G_1 and G_2 be two eulerian graphs with no vertices in common. Let v_1 be a vertex of G_1 and let v_2 be a vertex of G_2. Let G be a graph consisting of G_1 and G_2, together with the edge v_1v_2. What can be said about G?

4. (a) Show that if M is a traversable multigraph, then an eulerian multigraph can be constructed from M by the addition of a single edge.
   (b) Would the result in (a) be true if the two occurrences of the word "multigraph" were replaced by "graph"?
   (c) Would the result in (a) be true if the word "addition" were replaced by "deletion"?

*5. We know that a connected multigraph with no odd vertices contains an eulerian circuit, and a connected multigraph with exactly two odd vertices contains an eulerian trail. Try to determine what special property is exhibited by a connected multigraph with exactly four odd vertices. Try to prove your answer.

*6. Prove Theorem 3.3.

7. Figure 3.8 shows a diagram of the mystical town of Libb, with

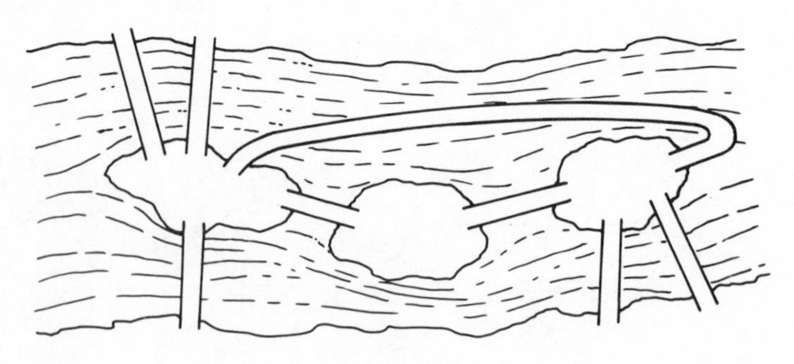

Figure 3.8

its three islands and many fabled bridges. Is it possible to walk through the town of Libb and cross each bridge once and only once? If so, show how such a walk can be made.

8. Figure 3.9 gives a (rather loosely drawn) map of the famous Snuff Islands off the coast of Zambesi. Boat routes for scenic tours are indicated on the map by dashed lines. Is it possible to make a round trip from Zambesi which follows each and every boat route exactly once?

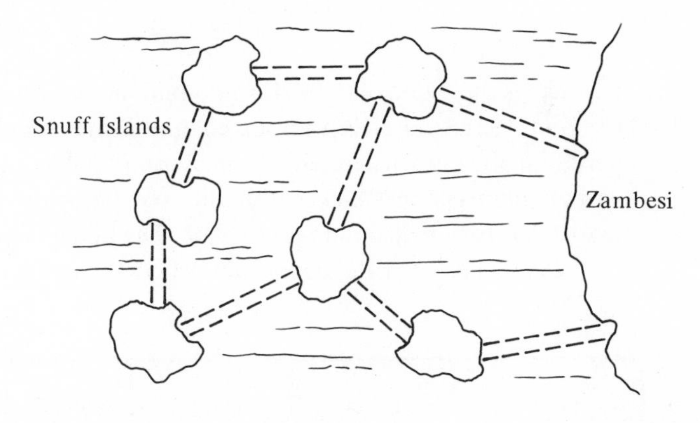

Figure 3.9

9. Suppose there is a group of four islands where a single boat route exists between each two islands. Is it possible to take a trip (not necessarily a round trip) that uses each boat route exactly once?

10. Figure 3.10 (page 64) shows a blueprint of a house. Can a person walk through each doorway of this house once and only once? If so, show how it can be done.

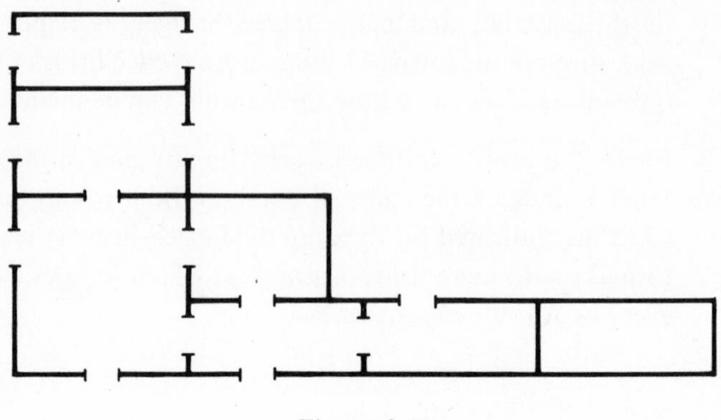

Figure 3.10

11. The scene is the estate of the well-to-do billionaire Count Van Diamond. He has just been murdered, and James Bomb, the internationally known detective, former notary public, current assistant manager of Chicken Delight, and part-time graph theorist, has been called in to investigate. The butler claims he saw the gardener enter the pool room (where the murder took

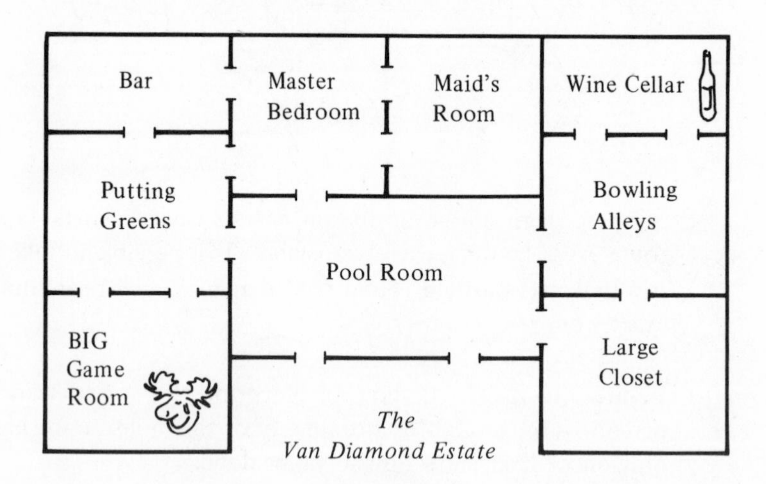

Figure 3.11

place) and then, shortly after, leave that room by the same door. The gardener, however, says that he cannot be the man whom the butler saw, for he entered the house, went through each door exactly once, and then left the house. James Bomb checks the floor plan (given in Figure 3.11). Within a matter of hours, he declares the case solved. Who killed the Count?

12. A letter carrier is responsible for delivering mail to houses on both sides of the streets shown in Figure 3.12. If the letter

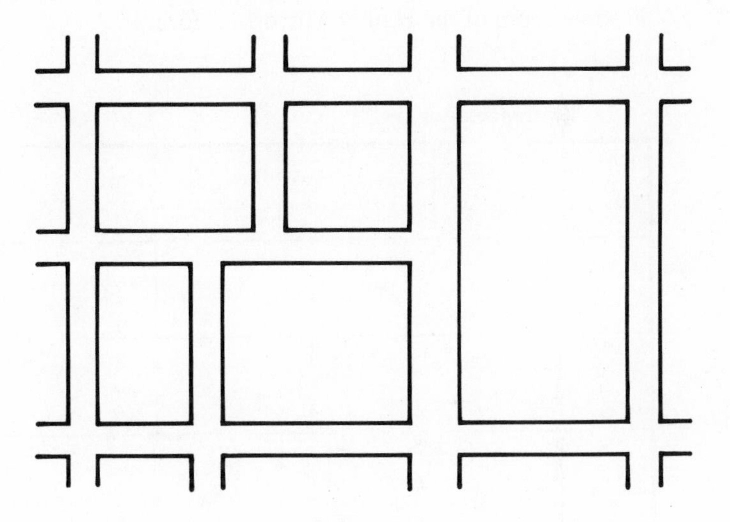

Figure 3.12

carrier does not keep crossing a street back and forth to get to houses on both sides of a street, it will be necessary for her to walk along a street at least twice, once on each side, to deliver the mail. Is it possible for the letter carrier to construct a round trip so that she walks on each side of every street exactly once?

*13. Does the solution of Exercise 3.12 depend significantly on the street diagram? Does this result suggest a theorem to you? If so, try to state the result and supply a proof.

14. Figure 3.13 is a diagram of the "Hall of Mirrors" at an amusement park. After each visitor passes through the entrance door, and through each door thereafter, the door automatically shuts and locks behind him. Assuming that you can eventually find your way out of any room if not all the doors in the room are locked, determine whether it is always possible to escape from the Hall of Mirrors, or whether you might become trapped in some room of the Hall of Mirrors . . . forever.

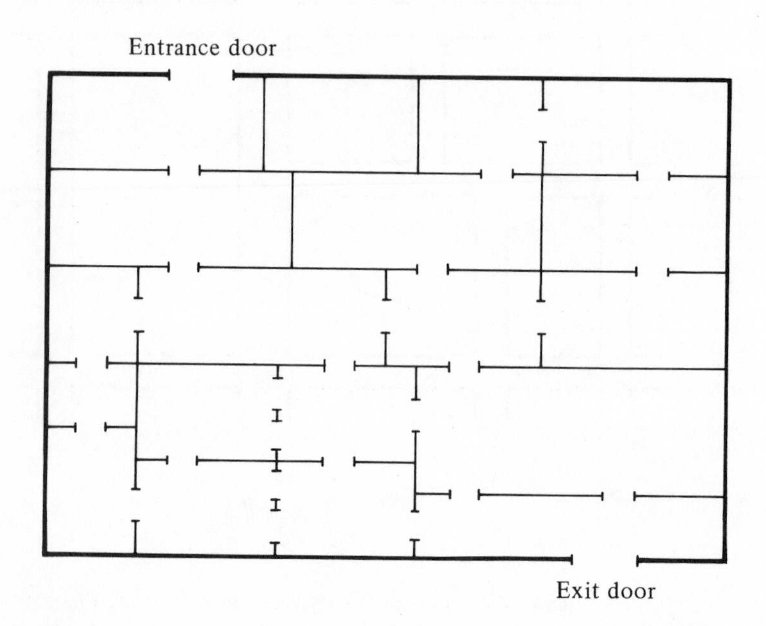

Figure 3.13

*15. Prove that a graph G is eulerian if and only if G is connected and its edge set can be partitioned into cycles.

3.2
The Salesman's Problem: An Introduction to Hamiltonian Graphs

 **The Salesman's Problem** 

SUPPOSE A SALESMAN'S TERRITORY includes several cities with highways connecting certain pairs of these cities. His job requires him to visit each city personally. Is it possible for him to schedule a round trip by car enabling him to visit each specified city exactly once?

We can represent this "transportation system" by a graph G whose vertices correspond to the cities, and such that two vertices are joined by an edge if and only if a highway connects the corresponding cities and does not pass through any other specified city. The solution to the problem depends on whether G has a cycle containing every vertex of G. (Note the difference between this problem and the problem in Section 3.1 of determining whether G has a circuit containing every vertex and edge of G.)

An important concept is suggested by this problem. We call a graph G *hamiltonian* if a cycle exists in G containing every vertex of G. A cycle containing all vertices of G is referred to as a *hamiltonian cycle*. Thus, a hamiltonian graph is a graph containing a hamiltonian cycle.

The graph G_1 of Figure 3.14 is hamiltonian, while G_2 is not hamiltonian. Graph G_1 is hamiltonian because it contains a hamiltonian cycle; for example, $u_1, u_2, u_5, u_4, u_3, u_1$ is a hamiltonian cycle. In order to show that G_2 is not hamiltonian, we give a proof by contradiction. Suppose, then, that G_2 is hamiltonian. Therefore, G_2 contains a hamiltonian cycle C. Now C contains every vertex of

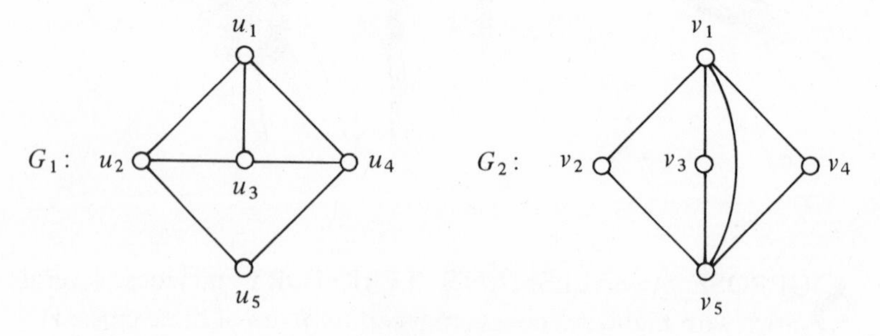

Figure 3.14

G_2; hence, C contains v_2, v_3, and v_4. Each of v_2, v_3, and v_4 has degree two, so C must contain the two edges incident with each of v_2, v_3, and v_4. This means, for example, that C contains all three edges v_1v_2, v_1v_3, and v_1v_4. However, any cycle can contain only two edges incident with a vertex on the cycle. Therefore, G_2 cannot contain a hamiltonian cycle, which contradicts the assumption that G_2 is hamiltonian.

It should now be apparent that the solution of any Salesman Problem depends on whether the associated graph is hamiltonian. Unfortunately, no one has yet found a convenient method for determining which graphs are hamiltonian. For the most part, each graph must be considered individually. However, some conditions have been developed which imply that the graph under consideration is necessarily hamiltonian. We present one of these results now.

Theorem 3.4

If G is a graph of order $p\,(\geq 3)$ such that $\deg v \geq p/2$ for every vertex v of G, then G is hamiltonian.

Proof

If G has order $p = 3$ and $\deg v \geq 3/2$ for every vertex v of G, then $\deg v = 2$ and $G = K_3$ (Figure 2.2, page 30). Therefore, the result is true for $p = 3$. We now assume $p \geq 4$. Among all the paths in G, let P be one of those paths with the greatest number of vertices. Suppose $P: u_1, u_2, \ldots, u_k$ is this path (see Figure 3.15).

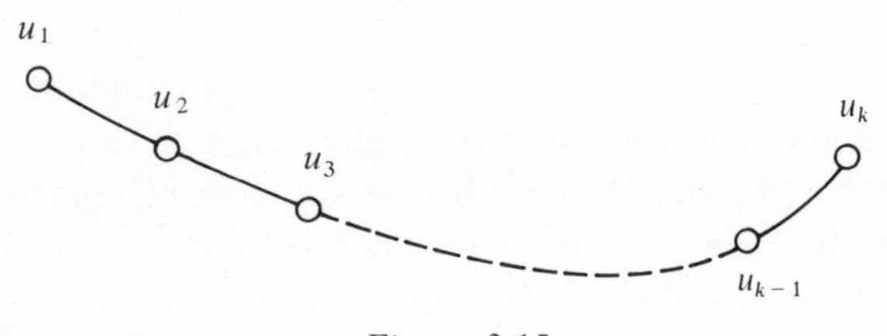

Figure 3.15

Since no path in G has more vertices than P, every vertex adjacent with u_1 must belong to P. Also, every vertex adjacent to u_k must belong to P. Since u_1 is adjacent to at least $p/2$ vertices, all on P, it follows that P must contain at least $1 + p/2$ vertices.

Now, there must be some vertex u_i on P, where $2 \leq i \leq k$, such that u_1 is adjacent to u_i and u_k is adjacent to u_{i-1}. If this were not the case, then for each vertex u_i adjacent to u_1, the vertex u_{i-1} would not be adjacent to u_k. However, since there are at least $p/2$ vertices u_i adjacent to u_1, there would be at least $p/2$ vertices u_{i-1} not adjacent to u_k. Therefore, $\deg u_k \leq (p - 1) - p/2 < p/2$, which is impossible since $\deg u_k \geq p/2$. Hence there is a vertex u_i on P such that u_1u_i and u_ku_{i-1} are both edges of G (see Figure 3.16). It now follows that there is a cycle $C: u_1, u_i, u_{i+1}, \ldots, u_k, u_{i-1}, u_{i-2}, \ldots, u_1$ containing all the vertices of P.

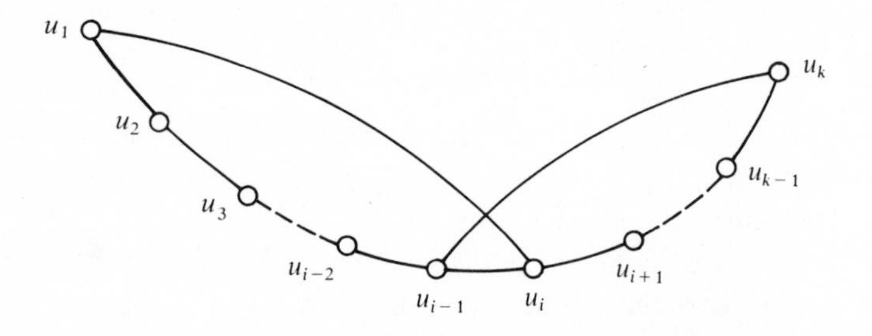

Figure 3.16

If all vertices of G belong to C, then C is a hamiltonian cycle and G is a hamiltonian graph. Suppose there is a vertex w of G that does not belong to C. Since C contains at least $1 + p/2$ vertices, fewer than $p/2$ vertices of G do not lie on C. Since $\deg w \geq p/2$, the vertex w must be adjacent to some vertex u_j of C. However, the edge

wu_j and the cycle C produce a path having one more vertex than P, which cannot occur since P has the greatest number of vertices. This contradiction implies that C contains all vertices of G, so that G is hamiltonian. ▮

While the condition that $\deg v \geq p/2$ for every vertex v of a graph G is sufficient for G to be hamiltonian, it is certainly not necessary. For example, G may be simply a cycle, in which case every vertex has degree two, but G is hamiltonian.

We should mention here how "hamiltonian" graphs got their name. It is said that the famous Irish mathematician Sir William Rowan Hamilton (1805–1865) invented a game which involved a regular solid dodecahedron (an object having 20 vertices, 30 edges, and 12 faces, this last property making it ideal as a desk-calendar paperweight). Hamilton labeled each vertex of the dodecahedron with the name of a well-known city. The object of the game was for the player to travel "Around the World" by determining a round trip which included all the cities exactly once, with the added

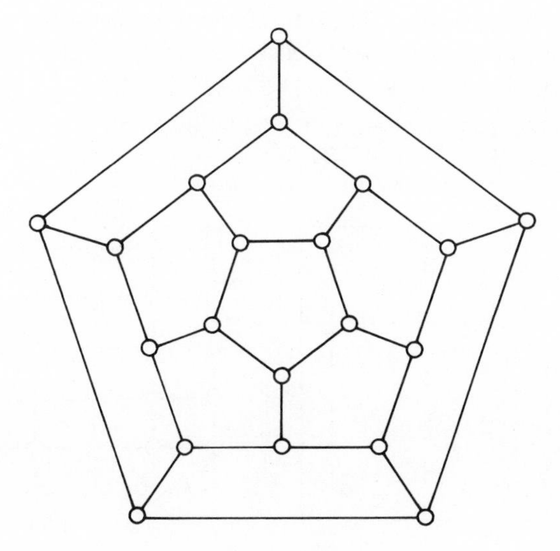

Figure 3.17

restriction that it is possible to travel from one city to another only if an edge exists between the corresponding vertices. Figure 3.17 shows the graph representing this problem, made up of the vertices and edges of the dodecahedron. Thus, the object of Hamilton's game is to find a "hamiltonian" cycle in the graph of the dodecahedron.

There are other puzzles whose solution can involve hamiltonian graphs. We consider one of these.

Example 3.3

Figure 3.18 shows a 6-by-6 maze (consisting of 36 squares). Is it possible to start at one of the squares, say the one in the upper left-hand corner, proceed to each square exactly once, and return to the starting square? This situation can be represented by the graph G in Figure 3.18. The vertices of G correspond to the squares of the maze and two vertices are joined by an edge if and only if we can move directly from one of the corresponding squares to the other. The solution to the maze depends on whether the graph G is hamiltonian, although the problem is probably easier to solve, in this case, without the aid of graphs. The answer to the question is affirmative, of course, with one solution shown by the dashed line in Figure 3.18.

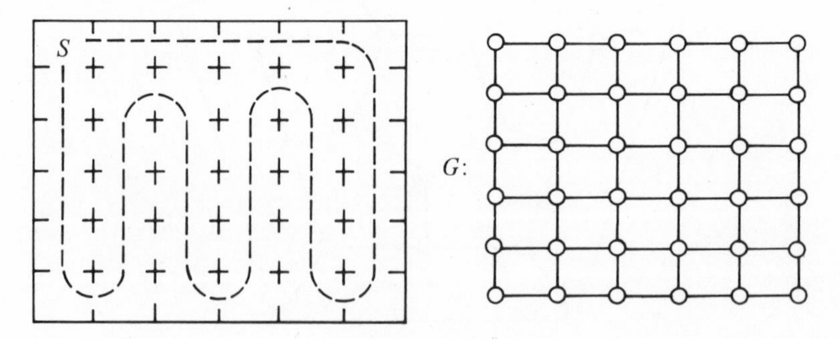

Figure 3.18

We conclude this section by mentioning a problem closely related to the Salesman's Problem. Suppose a traveling salesman is to visit certain cities on a trip and then return home. What is the *least expensive* such trip the salesman can make? This problem is called the ***Traveling Salesman Problem***, and has importance in the area of operations research. (Operations research applies mathematical theories and techniques to complex management problems.) The situation can be represented by an undirected network G whose vertices correspond to the various cities. Every two vertices are joined by an edge, and each edge is given a numerical value equal to the cost of traveling between the corresponding two cities. If we define the value of a hamiltonian cycle of G to be the sum of the values of its edges, then the problem is to find a hamiltonian cycle in G having minimum value. However, despite the simplicity of this statement of the Traveling Salesman Problem, no efficient solution has yet been found.

Problem Set 3.2

16. Determine which graphs in Figure 3.19 are hamiltonian.

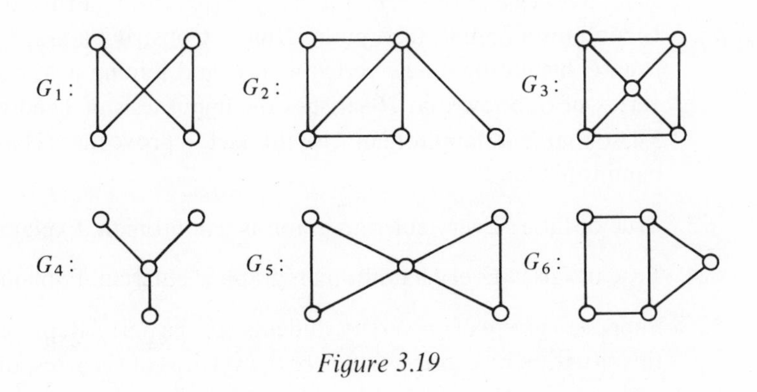

Figure 3.19

17. Give an example of a graph of order 10 which is
    (a) hamiltonian
    (b) not hamiltonian.

18. Show that the graph G of Figure 3.20 is not hamiltonian.

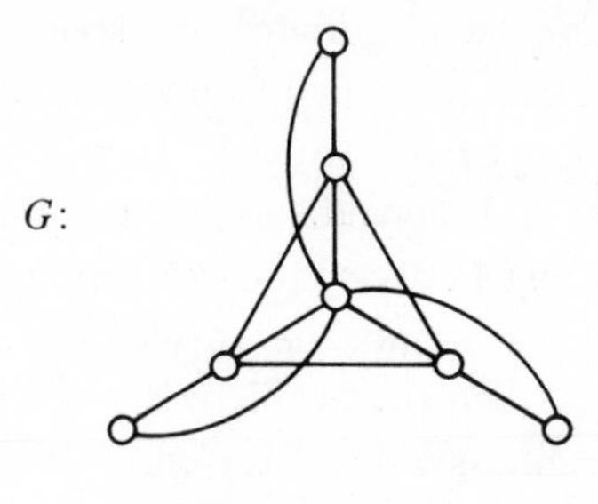

Figure 3.20

19. Show that the graph of a geometric cube is hamiltonian.

20. Learn what the geometric figure called an *icosahedron* is. Then show that the graph of an icosahedron is hamiltonian.

21. Show that Theorem 3.4 is no longer true if the number $p/2$ is replaced by $(p - 1)/2$.

*22. Let G be a graph. A path P in G is called a *hamiltonian path* if P contains every vertex of G. Let G be a graph of order $p\,(\geq 2)$ such that $\deg v \geq (p - 1)/2$ for every vertex v of G. Prove that G contains a hamiltonian path. (*Hint*: Construct a graph H from G by adding a new vertex w to G and joining w to each vertex of G. Show that H satisfies the hypothesis of Theorem 3.4, so that H is hamiltonian. Use this fact to prove that G has a hamiltonian path.)

23. True or false? Every eulerian graph is hamiltonian. Explain.

24. True or false? Every hamiltonian graph is eulerian. Explain.

*25. Suppose a group of university students are at a party. Represent this situation by a graph G, where the vertices of G correspond to the students at the party and two vertices of G are adjacent if and only if the corresponding man and woman have had a date together. If G is hamiltonian, prove that the number of men at the party equals the number of women at the party. (*Hint*: Label each vertex M or W according to whether the vertex corresponds to a man or a woman. Now use the fact that G possesses a hamiltonian cycle to establish the proof.)

26. Suppose we have a group of students (at least four) at a campus party, where the number of men at the party equals the number of women at the party. Suppose further that every man has had a date with every woman. Represent this situation by a graph G as in Exercise 25. Show that G is hamiltonian.

27. Determine whether the graph of Figure 3.21 is hamiltonian.

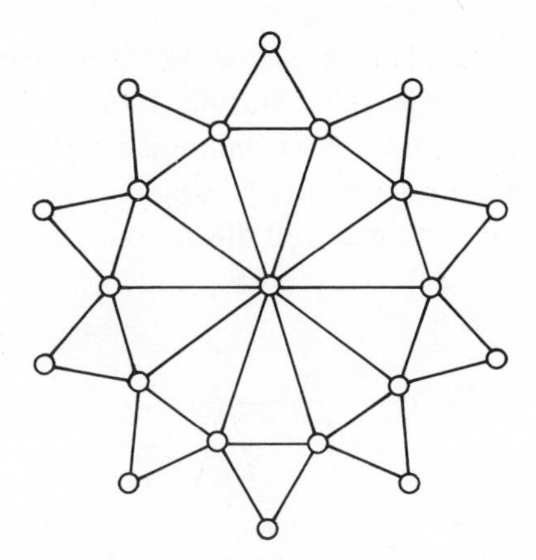

Figure 3.21

28. Referring to Example 3.2, show that in a 5-by-5 maze it is *not* possible to start at one of the squares, proceed to each square exactly once, and return to the starting square. (*Hint*: Color the squares alternately red and black.)

*29. Generalize Example 3.2 and Exercise 28 by finding exactly which pairs of positive integers m, n determine an m-by-n maze with the property that you may start at any square, proceed to each square exactly once, and return to the starting square.

30. Solve the Traveling Salesman Problem for the network shown in Figure 3.22.

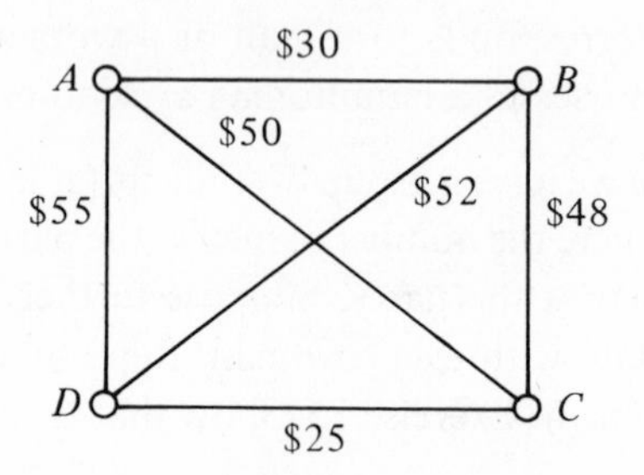

Figure 3.22

*31. Consider the Traveling Salesman Problem for the network shown in Figure 3.23. Starting from each of the five cities, determine the cost of a hamiltonian cycle obtained by always proceeding to the next city using the cheapest possible route. Show that in no case will this produce a hamiltonian cycle of least cost.

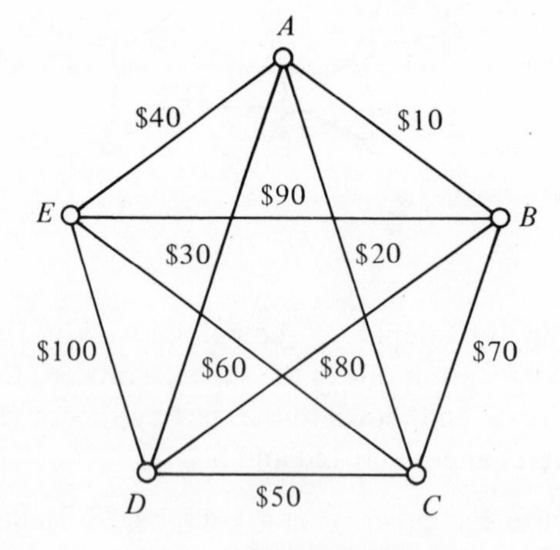

Figure 3.23

Suggestions for Further Reading

The student might enjoy reading Newman's [6] account of the Königsberg Bridge Problem, as well as Ball and Coxeter's [1] treatment of Hamilton's "Around the World" game. There are several expository articles written on hamiltonian graphs, including the rather elementary article [2]. An article dealing with eulerian graphs was written by Meyer [5]. You can find more information about the Traveling Salesman Problem by reading Flood [3]. Several variations of the transportation problems discussed in this chapter are mentioned in [4].

[1] W. W. R. Ball and H. S. M. Coxeter, *Mathematical Recreation and Essays.* Macmillan, New York (1947).

[2] G. Chartrand, S. F. Kapoor, and H. V. Kronk, "The many facets of hamiltonian graphs." *The Mathematics Student.* **41** (1973), pp. 327–336.

[3] M. Flood, "The travelling salesman problem." *Operations Research.* **4** (1956), pp. 61–75.

[4] L. Lesniak and G. Chartrand, "On traversing graphs." *Bulletin of the Iranian Mathematical Society.* **1** (1974), pp. 1–17.

[5] W. Meyer, "Garbage collecting, Sunday strolls, and soldering problems." *The Mathematics Teacher.* **65** (1972), pp. 307–309.

[6] J. Newman, "Leonhard Euler and the Königsberg Bridges." *Scientific American.* **189** (1953), pp. 66–70.

Chapter 4

Connection Problems

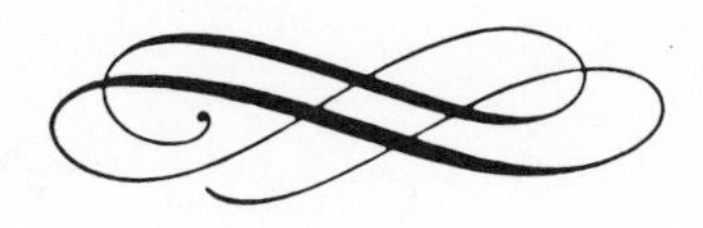

Numerous real-life situations involve proceeding as inexpensively or as quickly as possible from one location to another, or from one state to another. We consider two problems of this type, one involving graphs and the other involving digraphs.

4.1
The Minimal Connector Problem: An Introduction to Trees

The Minimal Connector Problem

SUPPOSE WE HAVE a collection of n cities, and we wish to construct a railroad system connecting them. Assume that we know the cost of building tracks between every two cities. Due to the current economic situation, we must construct the system as cheaply as possible—regardless of how inconvenient it turns out for prospective passengers. The question is: how should this railroad system be built?

The desired railway system can be represented by a graph G whose vertices correspond to the cities involved and in which two vertices are adjacent if a track runs between the two corresponding cities. It is essential, of course, that G be connected. However, it is also necessary that G not contain any cycles; for if G has a cycle and e is an edge on that cycle, then by Theorem 2.5, e is not a bridge and $G - e$ is connected. Hence, it would be possible to leave out a set of tracks between two cities and still have all cities connected, implying that the original railway system is not the cheapest. Thus, G must be connected but must contain no cycles.

Before solving the Minimal Connector Problem, we need to investigate in more detail the type of graph we have just encountered. Any connected graph that has no cycles is called a *tree.* As we already noted, every edge of a tree is a bridge. It is customary to define a graph without any cycles (be it connected or not) as a *forest.* Hence, each component of a forest is a tree. The word "tree" is used for these graphs because, when drawn, some of them may look like trees (see Figure 4.1).

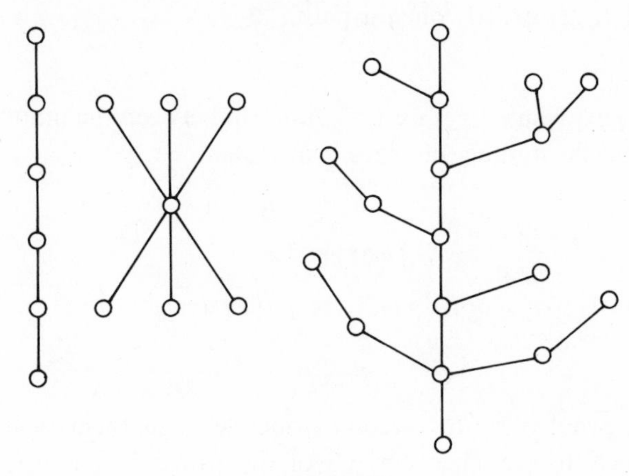

Figure 4.1

We now present a few interesting properties of trees.

Theorem 4.1

Let u and v be two vertices of a tree G. Then there is exactly one u-v path in G.

Proof

We know that there is at least one u-v path in G since every tree is connected. We show that G cannot have more than one u-v path. Suppose there are two u-v paths in G, say P_1 and P_2. Since the paths P_1 and P_2 are different, there must be a vertex w_1 (possibly $w_1 = u$) on both paths such that the vertex following w_1 on P_1 is not the same as the vertex following w_1 on P_2. Since P_1 and P_2 terminate at the vertex v, there is a first vertex after w_1, call it w_2, which P_1 and P_2 have in common (possibly

$w_2 = v$). However, that part of P_1 from w_1 to w_2 together with that part of P_2 from w_1 to w_2 form a cycle in G. This contradicts the fact that G has no cycles. Therefore, G has exactly one u-v path. ▮

The next result gives us a relationship between the number of vertices and the number of edges in any tree.

Theorem 4.2

If G is a tree of order p and size q, then $q = p - 1$.

Proof

The proof is by the second principle of mathematical induction (see Theorem A.8 of the Appendix.) Let S_n denote the statement: Every tree of order n has $n - 1$ edges. We prove that S_n is true for every positive integer n.

The statement S_1 is true since there is only one tree of order one and that tree has zero edges. Let $k > 1$ be a positive integer, and assume that the statements S_i, $1 \leq i < k$, are true. To prove that the statement S_k is true, we consider a tree T of order k. Let $e = uv$ be an edge of T. Since every edge of a tree is a bridge, the graph $T - e$ is disconnected. In fact, $T - e$ is a forest with two components, namely, a tree T_1 containing u, and a tree T_2 containing v. Suppose T_1 has order k_1 and T_2 has order k_2. We observe that $1 \leq k_1 < k$, $1 \leq k_2 < k$, and $k_1 + k_2 = k$. Since S_{k_1} and S_{k_2} are true statements, T_1 has $k_1 - 1$ edges and T_2 has $k_2 - 1$ edges. Therefore, T has size

$$(k_1 - 1) + (k_2 - 1) + 1 = k_1 + k_2 - 1 = k - 1.$$

Hence S_k is true, implying that S_n is true for every positive integer n. ▮

We now return to the Minimal Connector Problem. Suppose we have a connected undirected network G. (Thus, each edge of G has a value assigned to it.) We can construct a tree T such that T is a subgraph of G and T contains every vertex of G. Such a tree is called a *spanning tree* of G.

We wish to construct a particular spanning tree T of the connected undirected network G representing the possible railway systems of the Minimal Connector Problem. For the first edge e_1 of T, we select any edge of G of minimum value. As a second edge e_2 of T, we select any remaining edge of G having minimum value. For e_3 we choose any remaining edge of minimum value which does not form a cycle with the previously selected edges. We continue this procedure until we obtain a spanning tree T. A spanning tree arrived at in such a manner is called an *economy tree.*

It should be clear at this point that finding a solution to the Minimal Connector Problem is equivalent to determining, for a given connected undirected network G, a spanning tree T whose *value* (the sum of the values of its edges) is a minimum.

Theorem 4.3
Solution of the Minimal Connector Problem

Let G be a connected undirected network, and let T be an economy tree of G. Then T is a spanning tree whose value is a minimum.

Proof

For each edge e of G, let $f(e)$ denote the value of e. Furthermore, if H is a subgraph of G, then we define $f(H)$ to be $\sum f(e)$, where the sum is taken over all edges of H.

If the network G has order p, then an economy tree T of G has $p - 1$ edges. Let the edges of T be ordered

$e_1, e_2, \ldots, e_{p-1}$, as described earlier. The value $f(T)$ of T is then given by

$$f(T) = \sum_{i=1}^{p-1} f(e_i).$$

Let T_0 be a spanning tree of G having minimum value. We show that $f(T_0) \geq f(T)$, the desired result.

If the trees T_0 and T are not identical, then T has one or more edges which are not in T_0. Using the ordering on the edges of T, we let e_i, $1 \leq i \leq p - 1$, be the first edge of T not in T_0. We add the edge e_i to T_0, obtaining a graph G_0. Suppose $e_i = uv$. Then a u-v path P exists in T_0, and so P together with e_i produces a cycle C in G_0. Since T contains no cycles, there must be an edge e_0 in C which is not in T. The graph $T'_0 = G_0 - e_0$ is also a spanning tree of G, and

$$f(T'_0) = f(T_0) + f(e_i) - f(e_0).$$

However, we know that $f(T_0) \leq f(T'_0)$, so

$$f(T_0) - f(T'_0) \leq 0.$$

Since

$$f(T_0) - f(T'_0) = f(e_0) - f(e_i),$$

it follows that

$$f(e_0) \leq f(e_i).$$

However, by the manner in which T was constructed, e_i is an edge of smallest value which can be added to the edges $e_1, e_2, \ldots, e_{i-1}$ without producing a cycle. Also, if e_0 is added to the edges $e_1, e_2, \ldots, e_{i-1}$, no cycle is produced either. Therefore, $f(e_i) = f(e_0)$, so that $f(T'_0) = f(T_0)$.

We have shown in this fashion the existence of a spanning tree of minimum value, namely T'_0, such that the number of edges common to T'_0 and T exceeds the number of edges common to T_0 and T by one edge, namely e_i. By continuing this procedure, we finally arrive at a spanning tree with minimum value which is identical to T. Therefore, T has minimum value. ∎

Example 4.1

Suppose A, B, C, and D are four cities located at the corners of a square, one mile on a side. Assume that the cost of building a track between any two cities is proportional to the distance between them. Determine the least expensive railroad system connecting all four cities.

We can represent this situation by the undirected network G of Figure 4.2. The solution consists of determining an economy tree T of G. One such tree can be found by selecting the edges AB, BC, and CD.

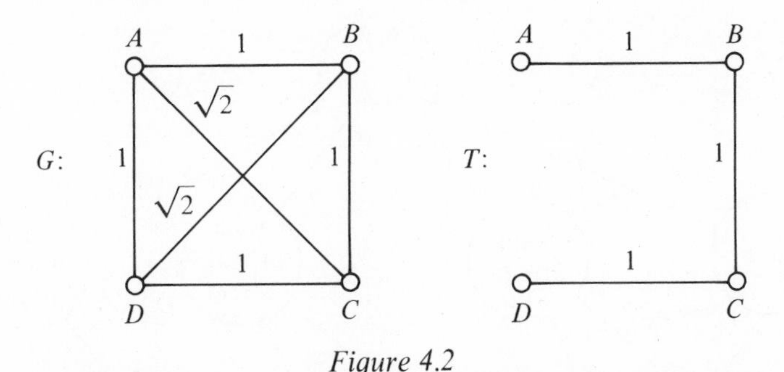

Figure 4.2

We consider a couple of other situations involving trees.

Perhaps one of the most familiar uses of trees occurs in family trees. Figure 4.3 shows the known ancestral family tree of international playboy Rupert Rupert.

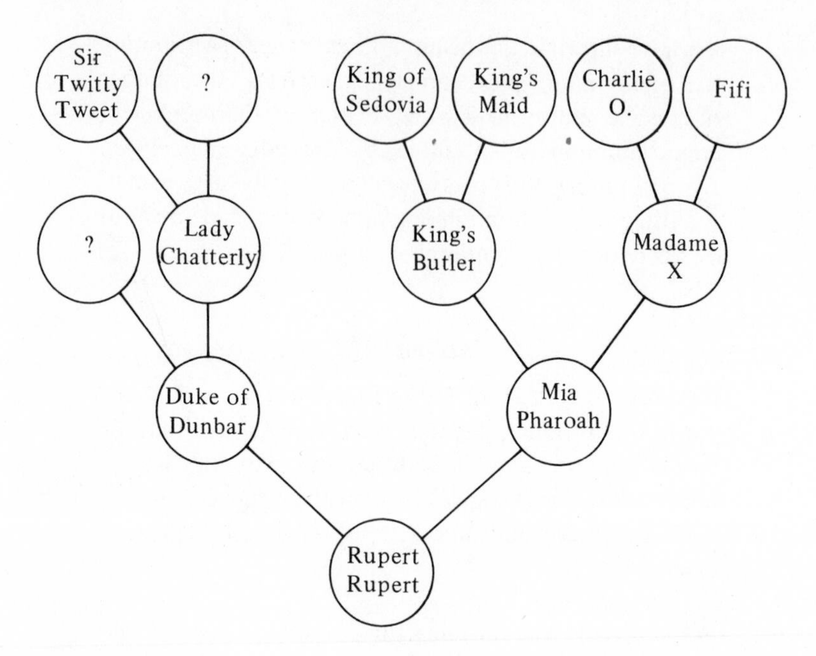

Figure 4.3

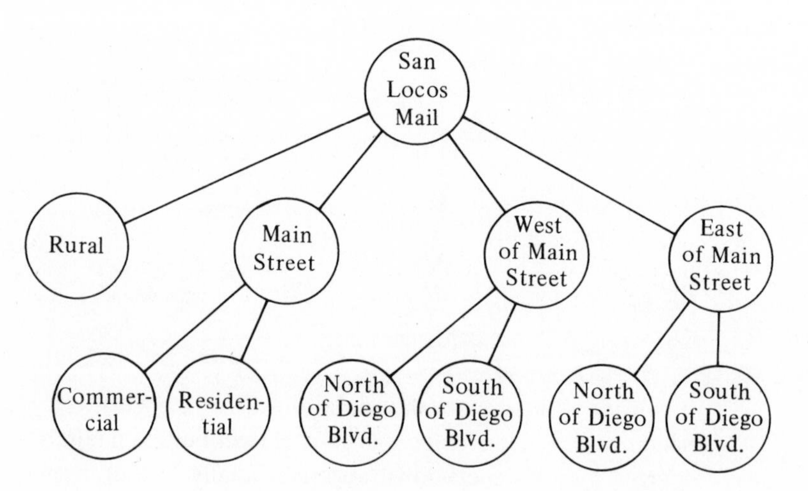

Figure 4.4

We also can describe sorting processes with the aid of trees. For example, in the small town of San Locos, incoming mail is sorted by hand. The mail is divided by districts and then more finely divided for delivery. Figure 4.4 shows one possible post office sorting plan.

Problem Set 4.1

1. How many different trees (pairwise non-isomorphic) are there of order:
   (a) two?
   (b) three?
   (c) four?
   (d) five?
   (e) six?
2. Determine the different (non-isomorphic) spanning trees of the graph G in Figure 4.5.

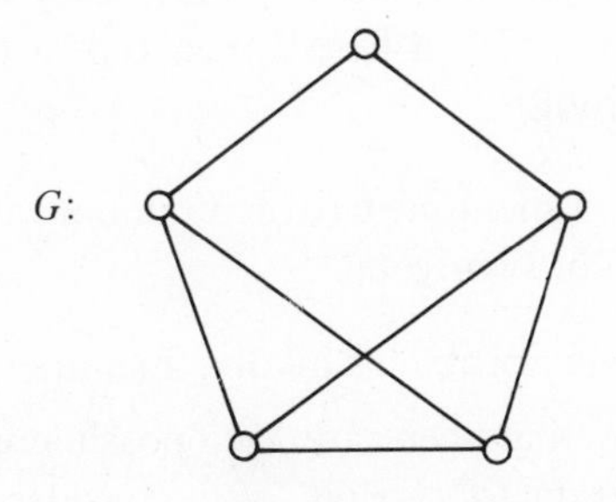

Figure 4.5

3. Is it possible for a tree to be a regular graph? Explain.

4. Let G be a graph possessing the property that for every two distinct vertices u and v of G, there exists exactly one u-v path. Prove that G is a tree.

5. By Theorem 4.1 and Exercise 4, it follows that trees are characterized by the fact that for every two distinct vertices u and v in a tree T, T contains exactly one u-v path (that is, trees are the only graphs with this property). Give an example of a graph having the property that for every two distinct vertices u and v, there exist exactly two u-v paths.

6. If a graph G of order p has size $q = p - 1$, must G be a tree? Explain.

*7. Let G be a forest of order p and size q having k components. Obtain an expression for q in terms of p and k.

*8. Prove that if G is a connected graph of order p and size q, then $q \geq p - 1$.

*9. If G is a connected graph of order p and size q such that $q = p - 1$, prove that G is a tree.

*10. Let G be a graph of order p and size q such that $q \geq p \geq 3$. Show that G must contain at least one cycle.

*11. Let G be a graph of order p and size q such that $q = p - 1$. Prove that if G is a forest, then G is a tree. (*Hint*: Try an indirect proof.)

12. Determine a spanning tree of minimum value in the undirected network G of Figure 4.6.

13. The following problems involve Example 4.1.

    (a) Show that there are four possible economy trees in G. What factors might you consider in deciding which economy tree to select?

    (b) A city E lies in the middle of the square determined by A, B, C, and D. The state chamber of commerce decides

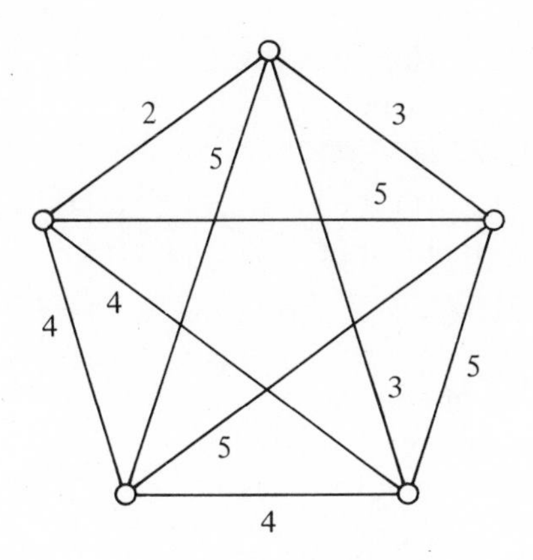

Figure 4.6

that all five cities should be connected by a railroad system. What is the least expensive system which can be constructed if:

(i) the system involving A, B, C, and D has already been constructed?

(ii) the system involving A, B, C, and D has not yet been constructed?

(c) Might there be four cities situated approximately as A, B, C, and D where the cost of constructing track between any two cities is not proportional to the distance between them? Explain how this might happen.

14. Agents A, B, C, D, E, F, G, and H are political conspirators in what has become known as the "Blottergate Affair." In order to coordinate their cover-up efforts, it is vital that each agent be able to communicate directly or indirectly with every other conspirator. Such communications involve a certain amount of risk to everyone. Below is a table of "risk factors" associated with direct communication between the indicated parties. All other direct communications are too likely to expose the

cover-up scheme. What is the least total risk involved in a connecting system?

Agent Pairs	*A* *B*	*A* C	*A* E	*A* F	*A* G	*B* C	*B* F	C D	C F	C G	C *H*	*D* E	*D* *H*	*E* *H*
Risk Factor	9	3	8	3	4	10	6	6	4	5	7	6	3	5

15. Draw as much of your family tree as you are familiar with.

*16. Given a tree T, what properties must T possess in order for it to be a family tree?

*17. Figure 4.7 presents a diagram of a prison for political dissidents. The prisoners have been divided into seven groups as shown. A spy plans to help all the prisoners escape by blowing up the gates in the prison walls. Due to the danger of this plan, he wants to destroy as few gates as possible and still allow all prisoners to escape. How many gates must be blasted to do this? Can a more general problem be solved? How is this related to trees?

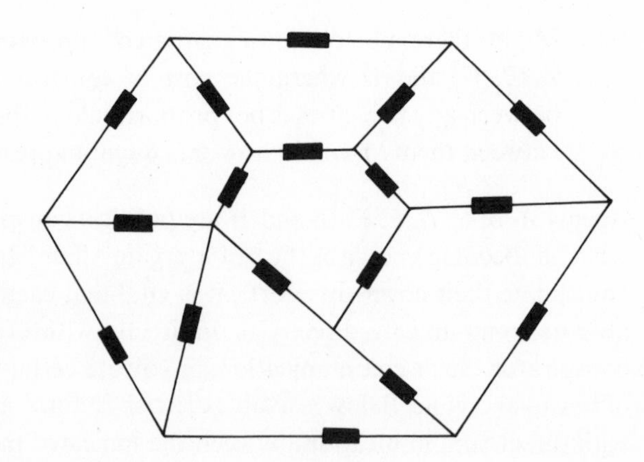

Figure 4.7

18. Trees can also be useful in categorizing the possible outcomes of certain events. For example, suppose football teams from Oakland and Kansas City play twice during the regular season. Assuming that a tie is unlikely, we give in Figure 4.8 the possible outcomes of the two games. Observe that there are two ways for Oakland (O) and Kansas City (K) to each win one of two games played, namely, the orders OK and KO. This equals exactly the $2! = 2$ permutations of the set $\{O, K\}$. With this example in mind, use a tree to aid you in listing all 24 permutations of the set $V = \{v_1, v_2, v_3, v_4\}$.

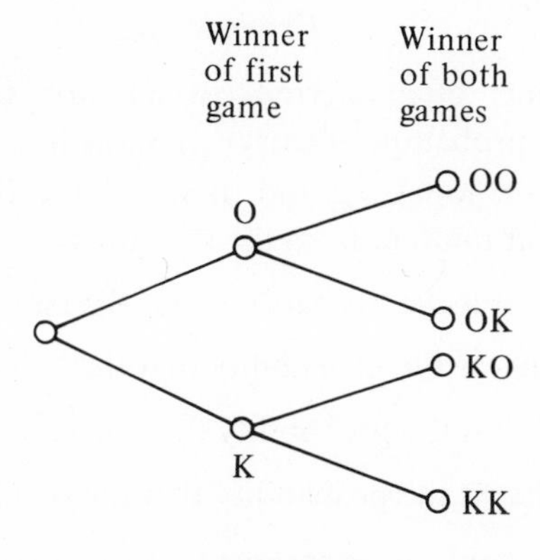

Figure 4.8

*4.2
Trees and Probability

We have just seen (in Exercise 18) how we can use trees in counting or listing possible outcomes of situations. Trees can also be helpful, in certain cases, in calculating probabilities. We illustrate this idea briefly in this section.

Probability theory is a vast and deep area of mathematics; however, we shall consider only a few elementary examples from this subject. Numerous books have been written on probability, and every finite mathematics book deals with this topic to some extent.

We shall not present a formal definition of probability. Intuitively, we say that probability deals with what are called "experiments" of various types and with possible outcomes, called "events," of the experiments. We consider a few examples to give the reader a taste of the subject.

Example 4.2

A container (often referred to as an "urn" by people who work in probability theory) contains five balls of equal size, two colored red and three colored blue. A ball is selected at random from the container.

(a) What is the probability that the ball is red?

(b) What is the probability that the ball is blue?

(c) What is the probability that the ball is red or blue?

(d) What is the probability that the ball is green?

In this example the experiment consists of selecting a ball. There are five possible outcomes and each outcome is equally likely to occur. Therefore, we say that (a) the probability of selecting a red ball is $2/5 = 0.4$, and (b) the probability of selecting a blue ball is $3/5 = 0.6$. Since every ball is red or blue, (c) the probability of selecting a red or blue ball is $5/5 = 1$. Since no ball is green, (d) the probability of selecting a green ball is $0/5 = 0$.

Example 4.3

A card is drawn at random from a shuffled, standard deck of 52 playing cards. What is the probability that the card is a jack?

Of the 52 cards, 4 are jacks; thus, the probability of drawing a jack is $4/52 = 1/13$.

Example 4.4

A fair coin is flipped. (A "fair" coin means that heads or tails are equally likely to appear if the coin is flipped.) What is the probability that heads comes up?

There are two possibilities: heads or tails. Since each outcome is equally likely, the probability of heads coming up is $1/2 = 0.5$.

Example 4.5

This is a continuation of the preceding example. A fair coin is flipped *twice*.

(a) What is the probability that heads comes up both times?

(b) What is the probability that tails comes up both times?

(c) What is the probability that heads and tails occur once each?

There is more than one way of answering these questions; however, we consider only one technique. The probability that heads occurs on the first flip is 0.5, as we saw in Example 4.4. The probability that heads occurs on the second flip does not depend on what happened on the first flip, so this probability is also 0.5. This "independence" implies that we obtain the answer to (a) by multiplying 0.5 and 0.5; that is, the probability that heads comes up both times is $(0.5)(0.5) = 0.25$. By the same kind of argument, the probability (b) that tails occurs both times is (0.5)(0.5) = 0.25. There are two ways that heads and tails can occur once each; namely, heads could occur on the first flip and tails on the second, or tails could occur on the first flip and heads on the second. This probability (c) is (0.5)(0.5) + (0.5)(0.5) = 0.5. The possibilities for the two flips and the corresponding probabilities can be given by means of a tree, as shown in Figure 4.9.

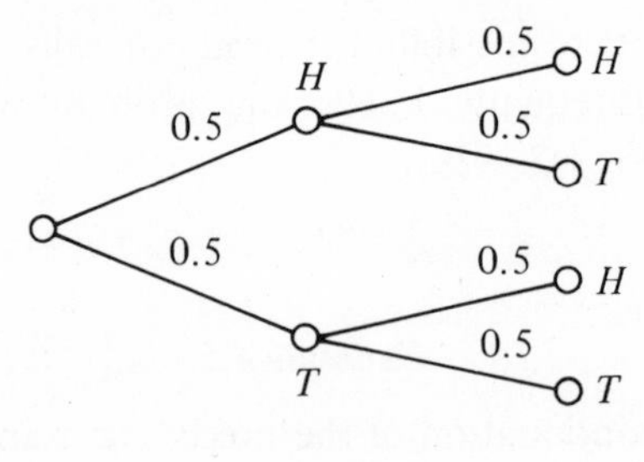

Figure 4.9

Reading from left to right, we see the four possibilities for the two flips of a coin: HH (heads followed by heads), HT, TH, and TT. The probability of each event is then obtained by multiplying the probabilities associated with the edges on the corresponding paths.

Example 4.6

We are faced with three containers, numbered 1, 2, and 3. Container 1 contains one black ball and two white balls. Container 2 has one black ball and three white balls, while Container 3 holds only one ball, colored black. Therefore, the three containers have a total of three black balls and five white balls. We select a container at random and then select a ball at random from the chosen container. Is it more likely that a black ball or a white ball is chosen?

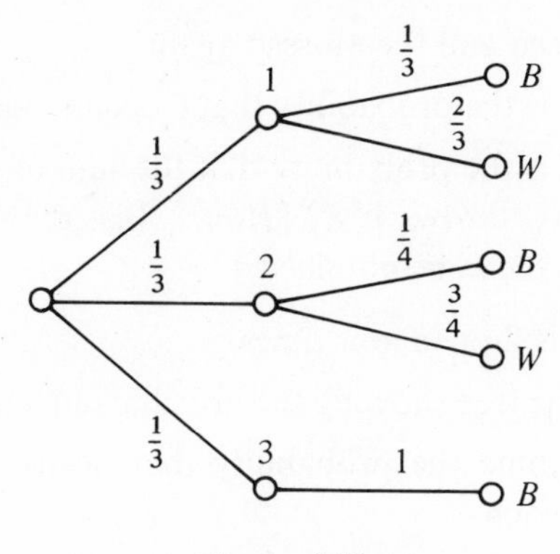

Figure 4.10

The situation is described by the tree of Figure 4.10. Since each container is equally likely to be selected, the probability that any particular one is chosen is 1/3. We calculate the probability that a white ball is selected. With the aid of the tree of Figure 4.10, we see that this probability is

$$\frac{1}{3}\cdot\frac{2}{3}+\frac{1}{3}\cdot\frac{3}{4}+\frac{1}{3}\cdot 0=\frac{2}{9}+\frac{1}{4}=\frac{17}{36}.$$

Since this probability is less than 0.5, we conclude that we are more likely to choose a black ball.

Problem Set 4.2

19. A die (a cube whose six faces show the numbers 1 through 6) is tossed.
    - (a) What is the probability that 6 is thrown?
    - (b) What is the probability that 1 is thrown?

20. A die is tossed and then tossed again.
    (a) What is the probability that 6 occurs both times?
    (b) What is the probability that the sum of the numbers on the two throws is 8? [*Hint*: List the various ways in which 8 can be obtained.]
21. A fair coin is flipped four times.
    (a) Draw a tree showing the possible outcomes.
    (b) Determine the probability that heads and tails occur twice each.
22. Two urns, labeled A and B, contain red and green balls of equal size. Urn A contains one red and two green balls; urn B holds three red and two green balls. An urn is selected at random and then a ball is selected at random from the chosen urn. Use a tree to determine which colored ball is more likely to be selected.
23. Two teams A and B have played several basketball games, with A winning two-thirds of the games. These teams are now involved in a play-off where the winner is the first team to win two games. Assuming the probability of A winning a game is 2/3, use a tree to determine the probability of A winning the play-off.

4.3
PERT and the Critical Path Method

The management sciences abound with situations requiring planning and scheduling. Many of the problems encountered are extremely complicated, and careful analysis of these problems commonly requires a sound background in a variety of mathematical subjects. However, with our background in graph theory, it is possible to acquire a feel for handling such problems.

In Section 4.1 we discussed paths and connection in graphs; in this section, we discuss paths and connection in digraphs (see Section 1.5). For vertices u and v of a digraph D, a *u-v path* in D is an alternating sequence of distinct vertices and arcs of D, beginning with u and ending with v, such that each arc is *incident from* the vertex immediately preceding it and *incident to* the vertex immediately following it. For example, in the digraph D of Figure 4.11, the sequence $v_1, e_1, v_3, e_3, v_4, e_5, v_5$ is a v_1-v_5 path. This path could be denoted more simply by v_1, v_3, v_4, v_5. Note that v_1, v_3, v_5 is not a path because of the direction of e_4. Similarly, v_1, v_3, v_4, v_6, v_5 is not a v_1-v_5 path.

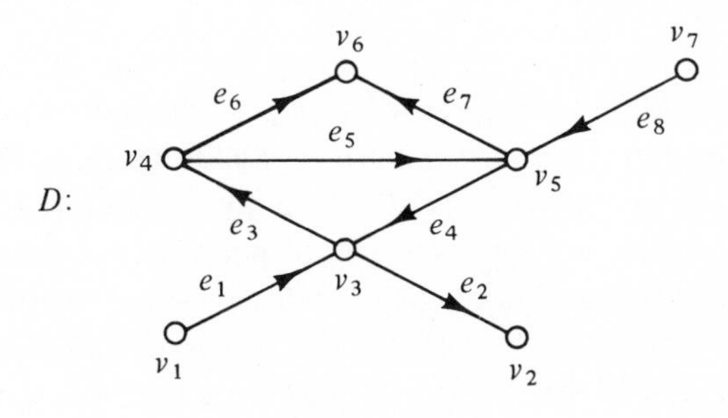

Figure 4.11

Suppose we have a project to complete. If the project is sufficiently complex, it is useful, if not essential, to thoroughly analyze it before making any attempt to begin work on the project. A common procedure is to divide the project into a number of smaller subprojects, which we call *activities*; that is, a project is composed of activities. Ordinarily, these activities are interrelated, and a certain activity may not be initiated until other activities have been completed. Let us consider a simple illustration before proceeding further.

Example 4.7

Suppose you have a rather spacious area in the basement of your home, and you would like to convert this space into a recreation room. You decide that you will panel the walls, install a ceiling, have the floor carpeted, and assemble a pool table. You may be rather handy at performing such "activities," but for the sake of the example, let us assume that you will have all these things done professionally. The activities, together with their estimated completion times, are:

	Activity	*Time* (hours)
P	Assemble pool table	4
F	Carpet the floor	6
W	Panel the walls	6
C	Install ceiling	8

It would be considerably simpler to assemble the pool table after the floor has been carpeted; hence we decide that F should precede P. With the exception of this restriction, any activities may be performed at the same time (although there is a possibility that the workers will get in each other's way). We can represent this situation by the "activity digraph" D of Figure 4.12. Each of the four activities is represented by a vertex, marked with the length of time necessary to complete the activity. We introduce two other vertices, labeled S and E, to indicate the start and end of the project. A time of zero is assigned to each of these "activities."

Although it is easy to see that the entire project could be completed in $4 + 6 + 6 + 8 = 24$ hours, it is not difficult to see that it could also be done in 10 hours. Specifically, after the carpeting is laid, the pool table could be assembled. These two activities take 10 hours, and while these two activities are in progress, the walls could be paneled *and* the ceiling could be installed.

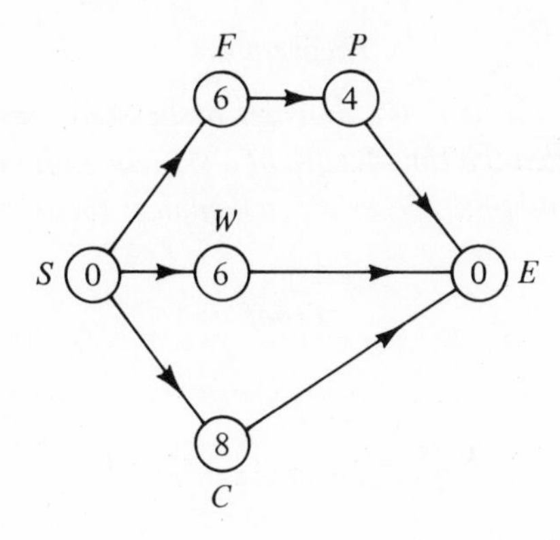

Figure 4.12

We now give a more formal definition of an activity digraph. Given a project composed of activities, we define a digraph D containing a vertex labeled S and a vertex labeled E, and such that the remaining vertices of D correspond to, and are labeled the same as, the activities. The vertex S is directed to a vertex v if the activity v may start without any other activity first being completed, and vertex w is directed to vertex E if no activity requires that w be completed before that activity begins. Furthermore, vertex x of D is directed to vertex y if and only if activity y may begin immediately after activity x is completed. Finally, we associate the number 0 with vertices S and E, and we associate the number t with vertex w if activity w requires t units of time for completion. The digraph D is called the *activity digraph* of the project.

Techniques referred to as *PERT* (Program Evaluation and Review Technique) and the *critical path method* make use of activity digraphs. A longest path (in units of time) in an activity digraph D is called a *critical path* of D. The activity digraph of Figure 4.12 has exactly one critical path, namely, S, F, P, E, and its *time-length* is 10 hours. It is no coincidence that the length of the critical path in this digraph equals the minimum time necessary to complete the project, as we now show.

Theorem 4.4

Let D be the activity digraph associated with a given project. Then the time-length of a critical path in D equals the minimum time necessary to complete the project.

Proof

Let

$$P : S = v_0, v_1, \ldots, v_n = E$$

be a critical path in D. Suppose the time-length of P is t. Since the activity v_i, $0 < i \leq n$, cannot begin until the activity v_{i-1} is completed, the time required to complete the entire project cannot be less than t; otherwise, not all of the activities of P could be completed.

It only remains for us to show that each activity of D may be performed during the t units of time needed to complete the activities of P. Let A be an arbitrary activity of the project. If A is an activity of P, then clearly A can be performed. Hence we assume A is not an activity of P.

Let Q be a path of maximum time-length containing A. Then Q must begin at start S and must terminate at end E. Hence the paths P and Q have at least two vertices in common, namely S and E. Therefore, these exists a subpath $Q' : w_0, w_1, \ldots, w_k$ of Q such that A belongs to Q', $w_0 = v_i$ and $w_k = v_j$ for $0 \leq i < j \leq n$, and none of the vertices $w_1, w_2, \ldots, w_{k-1}$ belong to P. Since P is a critical path, the time-length of the path

$$S = v_0, v_1, \ldots, v_i = w_0, w_1, w_2, \ldots, w_k = v_j, v_{j+1}, \ldots, v_n = E$$

is at most t. This implies that the activities $w_1, w_2, \ldots, w_{k-1}$ (including A) of Q may be performed while the activities

$v_{i+1}, v_{i+2}, \ldots, v_{j-1}$ of P are taking place. Since A was arbitrary, every activity of the project may be completed in t units of time. ▮

Example 4.8

A common example of the critical path method occurs in planning the construction of something rather complicated, such as a house. Let us assume that building a house can be described by the following activities:

	Activity	*Time* (days)
C	Clear land	1
F	Build foundation	3
U	Build upper structure	15
E_1	Electrical work	9
P	Plumbing work	5
E_2	Complete exterior work	12
I	Complete interior work	10
L	Landscaping	6

A possible activity digraph is indicated in Figure 4.13.

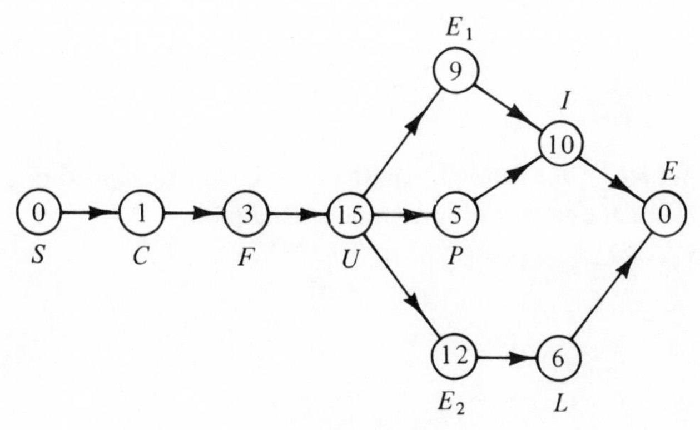

Figure 4.13

There is only one critical path in this case, namely S, C, F, U, E_1, I, E, and its time-length is 38 days. Thus, theoretically the house could be completed in 38 days.

As we mentioned at the beginning of this section, the discussion given here is meant only to present a flavor of the problems. In reality, this is a gross oversimplification. For example, if we say activity A_0 will take t_0 units of time, how do we really know this? Actually, t_0 may be only the average time it takes to perform activities similar to A_0. It is highly unlikely that the completion of A_0 will take exactly t_0 units, even after rounding off in some manner. In a particular case it might be valuable to know the minimum amount of time it ordinarily takes to complete A_0, as well as the maximum amount of time. If activity A_1 is to begin once A_0 is completed, we may wish to have our workers present and ready to begin A_1 just as A_0 concludes. However, in all probability this must be set up according to some prearranged schedule. It is quite possible that when the workers for A_1 arrive, either A_0 has not yet been completed, or A_0 was completed earlier and time has been wasted. In either case, it is likely that money also has been wasted.

Problem Set 4.3

24. Which of the following paths exist in the digraph D of Figure 4.11? If one exists, determine such a path.
    (a) v_7-v_6 path
    (b) v_1-v_7 path
    (c) v_1-v_6 path
    (d) v_4-v_2 path
    (e) v_7-v_1 path

25. Assign "times" to the vertices of the activity digraph D of Figure 4.14 and use them to determine a critical path in D.

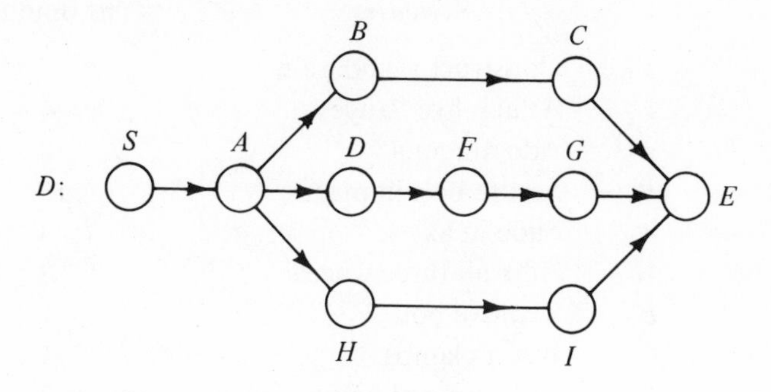

Figure 4.14

26. Explain why the digraph of Figure 4.15 cannot be an activity digraph.

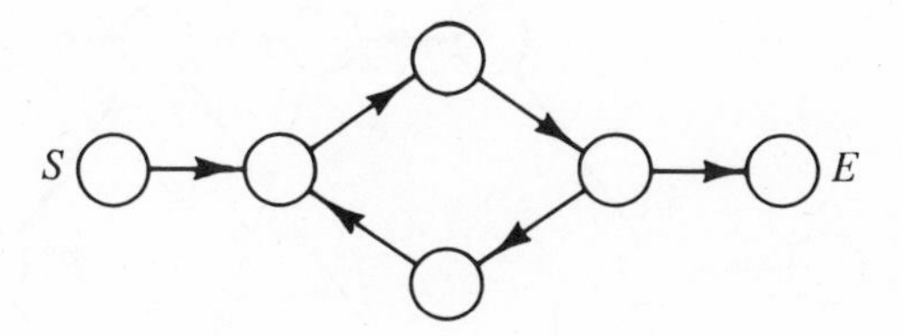

Figure 4.15

27. Explain why the digraph of Figure 4.16 cannot be an activity digraph.

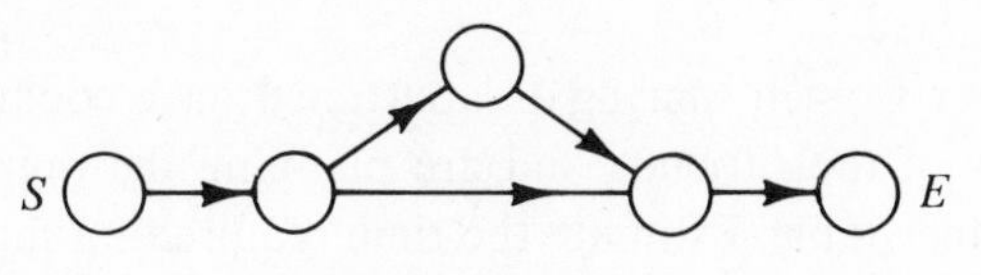

Figure 4.16

28. Suppose we wish to construct a widget. The usual way of doing this is to divide this project into the following activities:

	Activities	*Time* (minutes)
F_1	Construct foundation	5
F_2	Attach five flanges	4
A_1	Add antenna	8
B_1	Secure bolt in place	15
A_2	Slide in axles	3
W	Affix all three wheels	2
B_2	Remove bolt	1
C	Insert clamps	4
P	Two coats of paint	2

Assume the activity digraph is given in Figure 4.17. Determine all critical paths and their time-lengths.

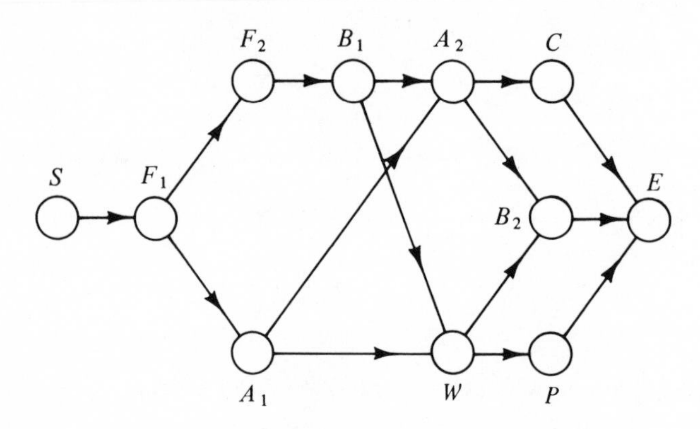

Figure 4.17

29. Four women sharing an apartment have cooked dinner for a few of their friends and are planning the preparation of the dinner table. Estimate the time required (in minutes) to complete each of the following activities, draw a possible activity

digraph, and determine the minimum time needed to complete the project.

Activities

C	Put tablecloth on table
N	Fold napkins and place on table
D	Place dishes and silverware on table
W_1	Put water and ice in water glasses
W_2	Pour wine into wine glasses
F	Place plates of food on table

30. Think of a project of your own and analyze it by means of an activity digraph.

31. Referring to Example 4.8, suppose you are the contractor. The workers who are to build the foundation say it will take them from 2 to 4 days to complete the job. What are the advantages and disadvantages of having the workers who are to build the upper structure report on the job 2, 3, or 4 days after the foundation is started?

*32. Characterize those digraphs which are activity digraphs, i.e., determine a necessary and sufficient condition for a digraph to be an activity digraph, and give a proof of this result.

Suggestions for Further Reading

For other versions of "The Minimal Connector Problem," the reader should refer to Ore [6] and Wilson [7]. Articles pertaining to trees and their properties include those by Anderson and Harary [1] and Harary and Manvel [3]. Much has been written on probability and trees; one interesting and readable paper is by Zimmerman [8].

A good discussion of the critical path method is given in Malkevitch and Meyer [4]. More advanced treatments of this subject may be found in Busacker and Saaty [2] and in Marshall [5].

[1] S. S. Anderson and F. Harary, "Trees and unicyclic graphs." *The Mathematics Teacher*, **60** (1967), pp. 345–348.

[2] R. G. Busacker and T. L. Saaty, *Finite Graphs and Networks.* McGraw-Hill, New York (1965).

[3] F. Harary and B. Manvel, "Trees." *Scripta Mathematica*, **28** (1970), pp. 327–333.

[4] J. Malkevitch and W. Meyer, *Graphs, Models, and Finite Mathematics.* Prentice-Hall, Englewood Cliffs, N.J. (1974).

[5] C. W. Marshall, *Applied Graph Theory.* Wiley-Interscience, New York (1971).

[6] O. Ore, *Graphs and Their Uses.* Random House, New York (1963).

[7] R. J. Wilson, *Introduction to Graph Theory.* Academic Press, New York (1972).

[8] M. Zimmerman, "Matrix multiplication as an application of the principle of combinatorial analysis." *Pi Mu Epsilon Journal*, **6** (1975), pp. 166–175.

Chapter 5

Party Problems

Numerous situations with graph theory overtones are expressible in terms of gatherings of people. For the purpose of adding a little spice to the problems, we may further assume, without loss of generality, that these people are attending a party. For example, suppose several people are at a party, and certain of these people are friends of one another. Is it possible to seat everyone around a large circular table so that each guest is sitting next to two of his or her friends?

Of course, it is impossible to answer the question with only the information given. However, the situation can be modeled by a graph G whose vertices represent the people and where two vertices are adjacent if and only if the corresponding individuals are friends of each other. Then the stated question has an affirmative answer if and only if G is hamiltonian. In this chapter we consider some other problems described in terms of parties.

Chapter 5
Party Problems

5.1
The Problem of the Eccentric Hosts: An Introduction to Ramsey Numbers

A man and his wife are planning a dinner party, but they each have certain peculiarities. The man likes to have everyone at the dinner table know one another, for he feels this creates more harmony during the meal. On the other hand, his wife thinks that no two people at the dinner table should know each other, for she believes one important aspect of dinner parties is the opportunity it gives people for making new acquaintances. Despite this difference of opinion between husband and wife, they are a happily married couple, and both will be happy if either's wishes are fulfilled. Hence, two tables A and B are set. At table A, the man's table, all guests are to know one another; while at table B, the wife's table, no two guests are to know each other. We may now state our problem.

The Problem of the Eccentric Hosts

SUPPOSE TABLE A is to seat m people and table B is to seat n people. All guests at table A are to know one another, while no two people seated at table B are to know each other. What is the least number of people that may attend a dinner party so that at least one of the two tables can be filled according to these rules?

For each collection of p people present at the party, we can represent their situation by a graph G of order p whose vertices correspond to the people, and such that two vertices of G are adjacent if and only if the corresponding people know each other. In order to state the Problem of the Eccentric Hosts in mathematical terms, however, a few additional definitions are convenient.

The *complement* $\bar{G}$ of a graph G is a graph having the same vertex set as G, but with the property that two vertices of $\bar{G}$ are adjacent if and only if the same two vertices of G are *not* adjacent. Figure 5.1 shows a graph G and its complement $\bar{G}$.

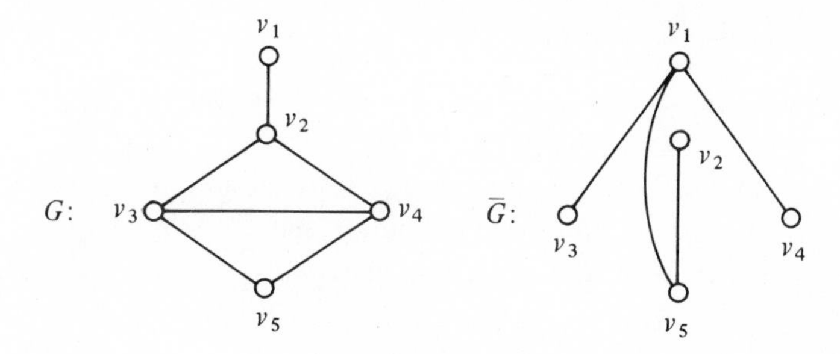

Figure 5.1

Let F_1 and F_2 be any two graphs. We define the *Ramsey number* $r(F_1, F_2)$ to be the least integer p such that for *every* graph G of order p, either G contains F_1 as a subgraph or $\bar{G}$ contains F_2

as a subgraph. These numbers are named after Frank Ramsey, who proved that for every two complete graphs K_m and K_n, the Ramsey number $r(K_m, K_n)$ exists. This, in turn, implies that $r(F_1, F_2)$ exists for every two graphs F_1 and F_2.

The following result is rather elementary, yet important.

Theorem 5.1

For every two graphs F_1 and F_2,

$$r(F_1, F_2) = r(F_2, F_1).$$

Proof

Let $r(F_1, F_2) = p$ and $r(F_2, F_1) = p_0$. We show that $p = p_0$. Let H be any graph of order p, and consider $\overline{H}$, also of order p. Since $r(F_1, F_2) = p$, either $\overline{H}$ contains F_1 as a subgraph or $\overline{\overline{H}} = H$ contains F_2 as a subgraph. Hence, for every graph H of order p, either H contains F_2 as a subgraph or $\overline{H}$ contains F_1 as a subgraph. Because $r(F_2, F_1) = p_0$, the number p cannot be smaller than p_0, so $p \geq p_0$.

Now suppose G is an arbitrary graph of order p_0. The graph $\overline{G}$, also of order p_0, either contains F_2 as a subgraph or $\overline{\overline{G}} = G$ contains F_1 as a subgraph. Since $r(F_1, F_2) = p$, however, it follows that $p_0 \geq p$, which, together with the preceding inequality, implies that $p = p_0$. ▮

The Problem of the Eccentric Hosts is equivalent to determining the Ramsey numbers $r(K_m, K_n)$. Unfortunately, for $m \geq 3$ and $n \geq 3$, very few Ramsey numbers have been found; hence, the Problem of the Eccentric Hosts is, in general, unsolved.

It is often convenient to investigate Ramsey numbers in another way. Consider a graph G of order p and its complement $\bar{G}$. The edge sets $E(G)$ and $E(\bar{G})$ partition the edge set of K_p; that is, every edge of K_p belongs either to G or to $\bar{G}$. If we take an edge of K_p belonging to G and color it red, and take an edge belonging to $\bar{G}$ and color it blue, then we may give an equivalent definition of Ramsey numbers. For graphs F_1 and F_2, the Ramsey number $r(F_1, F_2)$ is defined as the least integer p such that if every edge of K_p is colored red or blue (in any manner whatsoever), then there exists either a red F_1 (a subgraph F_1 with all edges colored red) or a blue F_2.

It follows from the definition(s) that if either F_1 or F_2 has order 1, then $r(F_1, F_2) = 1$. As further illustrations, we evaluate $r(K_3, K_2)$ and $r(K_3, K_3)$.

Theorem 5.2

$$r(K_3, K_2) = 3.$$

Proof

First we observe that $r(K_3, K_2) > 2$, since if the one edge of K_2 is colored red, then K_2 contains neither a red K_3 nor a blue K_2. This shows it is not true that for every graph of order two whose edge is colored red or blue, there exists either a red K_3 or a blue K_2.

We can verify, however, that if the edges of K_3 are colored red or blue, then there exists a red K_3 or a blue K_2. If all edges are colored red, we have a red K_3; if not, at least one edge is colored blue, producing a blue K_2. Therefore, $r(K_3, K_2) = 3$. ▮

One of the best known Ramsey numbers is $r(K_3, K_3)$.

Theorem 5.3

$$r(K_3, K_3) = 6.$$

Proof

First we observe that $r(K_3, K_3) > 5$ by noting that neither the graph H of Figure 5.2 nor its complement $\overline{H}$ contains K_3 as a subgraph.

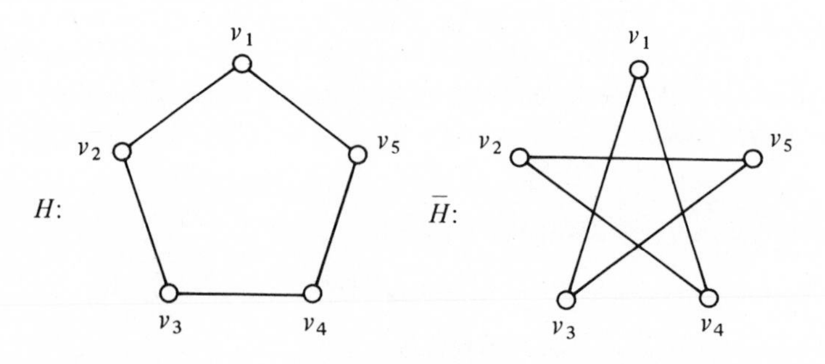

Figure 5.2

In order to show that $r(K_3, K_3) = 6$, it suffices to prove that if the edges of K_6 are colored red or blue, there is a red K_3 or a blue K_3. Consider some vertex of K_6, say v_1. Since v_1 is incident with five edges, at least three of the five edges must be colored the same, say red. (The argument is essentially the same if the majority of the edges incident with v_1 are colored blue.) Suppose v_1v_2, v_1v_3, and v_1v_4 are red edges. If any of the edges v_2v_3, v_2v_4, and v_3v_4 are red, we have a red K_3; if not, these three edges form a blue K_3. Hence, $r(K_3, K_3) = 6$. ▮

From Theorem 5.3, it follows that the eccentric hosts must have at least six people present at their dinner party in order to be absolutely certain that either three mutual acquaintances may be seated at the husband's table or three mutual strangers may be seated at the wife's table.

Problem Set 5.1

1. Figure 5.2 shows a cycle of order 5 and its complement. Draw a cycle of order 4 and its complement. Do the same for a cycle of order 6.

2. Let G be a graph of order p and size q. Determine the order and size of $\bar{G}$.

3. Let G be a graph whose vertices have degrees $d_1, d_2, \ldots, d_{10}$. Determine the degrees of the vertices of $\bar{G}$.

*4. Let G be a disconnected graph. Prove that $\bar{G}$ is connected.

5. Give an example of a connected graph whose complement also is connected.

*6. Let G be a regular graph of even order $p \geq 4$. Prove that at least one of G and $\bar{G}$ is hamiltonian.

7. A graph G is *self-complementary* if G and $\bar{G}$ are isomorphic. Figure 5.2 shows that a cycle of order 5 is self-complementary. Also, K_1 is self-complementary.

   (a) Give an example of a self-complementary graph of order 4.

   (b) Show that there are no self-complementary graphs of order 2 or 3.

*8. See the definition of "self-complementary" in Exercise 7.

   (a) Show that if a graph G has order p, and p is of the form $p = 4n + 2$ (i.e., $p = 2, 6, 10, \ldots$) or p is of the form $p = 4n + 3$ (i.e., $p = 3, 7, 11, \ldots$), then G is *not* self-complementary.

(b) Use part (a) as a hint to find a necessary condition for a graph G of order p to be self-complementary.

(c) Find a necessary and sufficient condition for a positive integer p to be the order of some self-complementary graph.

9. As we mentioned, Ramsey proved that for every two complete graphs K_m and K_n, $r(K_m, K_n)$ exists. What exactly does this mean?

10. Let F_1 be a graph of order m and F_2 be a graph of order n. Explain why $r(F_1, F_2) \leq r(K_m, K_n)$.

11. Suppose that the edges of K_7 are colored red and blue. Why does it follow that K_7 must contain a red K_3 or a blue K_3?

*12. Let G be a graph of order $r(K_m, K_n) - 1$, where $m, n \geq 2$. Prove that either G contains K_{m-1} as a subgraph or $\bar{G}$ contains K_{n-1} as a subgraph.

*13. Let P_n denote a path with n vertices (and $n - 1$ edges).

(a) Determine $r(P_3, P_3)$.

(b) Determine $r(P_3, K_3)$.

*14. (a) Generalize the concept of Ramsey numbers to define $r(F_1, F_2, F_3)$ and, more generally, $r(F_1, F_2, \ldots, F_n)$. [*Hint*: Consider the second definition discussed for $r(F_1, F_2)$.]

(b) Show that $r(K_2, K_2, K_2) = 2$.

(c) Determine $r(K_2, K_2, K_3)$.

(d) Determine $r(K_2, K_3, K_3)$.

15. Suppose we have a group of people attending a party, and every two people are either acquaintances or strangers to each other. How many people must be present at the party to guarantee the presence of either three mutual acquaintances or three mutual strangers? Explain.

*16. Suppose we have a group of people attending a party, and every two people are either acquainted or unknown to each other.

How many people must be present at the party for some person to be an acquaintance of at least three people or for some person to be a stranger to at least three people?

5.2
The Dancing Problem: An Introduction to Matching

Once again we are planning a party. This time it is to be quite an affair, and no expense is being spared. Indeed, even a band (Cicero Fudge and His Sweet Notes) has been hired, and there will be dancing. During each musical number, we would like as many guests dancing (with a partner) as possible. Of course, if we wish to have all guests dancing at one time, then we must invite an equal number of men and women; however, this may be neither possible nor desirable. Let us assume that there are at least as many men at the party as there are women. Then we would like every woman dancing with a male partner during each musical selection. This may not be easy to accomplish, however, for it may happen that each woman will only dance with certain of the men, and vice versa. For example, a man may very well be willing to dance with his wife, but this idea might be totally repugnant to his wife. Hence our goal is not only to have as many guests dancing as possible, but also to match all dancers with compatible partners. We now state our problem.

The Dancing Problem

SUPPOSE WE HAVE A GROUP of men and women at a party, with at least as many men present as there are women. Under what conditions is it possible to have all women dance with equally many men so that each dancing couple is compatible?

Chapter 5
Party Problems

As before, we explore this problem by setting up a mathematical model using graphs. Let G be a graph whose vertex set represents the people at the party, and such that two vertices are adjacent if and only if the corresponding people are compatible dancing partners. It is possible for all women to be dancing at the same time if and only if G contains a 1-regular subgraph F such that the number of edges in F equals the number of women.

A few new graphical definitions will make this problem easier to discuss. A graph G is called *bipartite* if it is possible to partition the vertex set of G into two subsets, say V_1 and V_2, so that every edge of G joins a vertex of V_1 with a vertex of V_2, and no vertex joins another vertex of its own set. In Figure 5.3 we have redrawn a graph G to illustrate its bipartite property. Thus, for this graph, we could let $V_1 = \{v_1, v_3, v_5, v_7\}$ and $V_2 = \{v_2, v_4, v_6, v_8, v_9\}$.

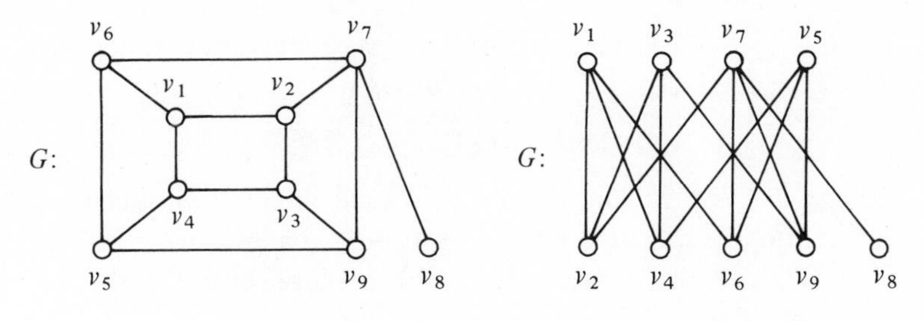

Figure 5.3

Let G be a bipartite graph whose vertex set is partitioned into subsets V_1 and V_2, as just described. Let U_1 be a subset of V_1. We say that U_1 is *matched* to a subset U_2 of V_2 if G contains a 1-regular subgraph F whose vertex set is $U_1 \cup U_2$. If U_1 is matched to U_2, then we must have $|U_1| = |U_2|$. The subgraph F is referred to as a *matching*, for it matches (or pairs off) one set of vertices, namely U_1, with another set of vertices, namely U_2.

We illustrate this concept with the bipartite graph G of Figure 5.4. Here we let $V_1 = \{v_1, v_2, v_3, v_4\}$, $V_2 = \{w_1, w_2, w_3, w_4, w_5\}$, and $U_1 = \{v_1, v_3, v_4\}$. If we let $U_2 = \{w_1, w_2, w_5\}$, then we see that G contains the 1-regular subgraph F with vertex set $U_1 \cup U_2$. Hence, U_1 is matched to U_2. Note that V_1 itself can be matched to a subset of V_2.

Again, let us assume G to be a bipartite graph with its vertex set partitioned into V_1 and V_2. If $U_1 \subseteq V_1$, then we denote by U_1^*

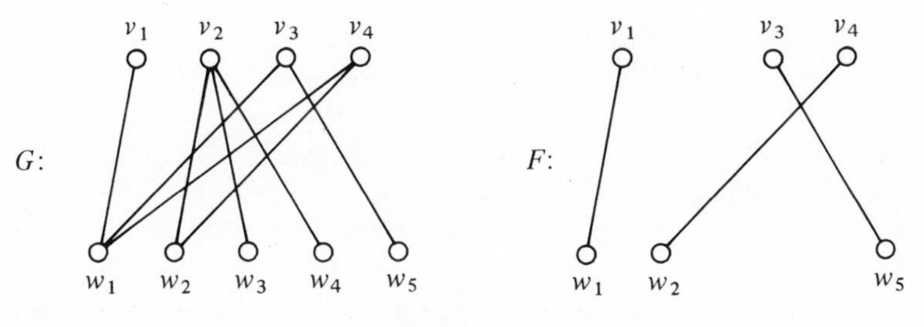

Figure 5.4

those vertices of V_2 adjacent with at least one vertex in U_1. The *deficiency* $\text{def}(U_1)$ of U_1 in G is defined by

$$\text{def}(U_1) = |U_1| - |U_1^*|.$$

The set U_1 is then said to be *nondeficient* in G if no (nonempty) subset of U_1 has positive deficiency. We note that positive deficiency for a subset W_1 of U_1 means that there are more vertices in W_1 than there are vertices adjacent to the elements in W_1; hence, it would be hopeless to attempt matching W_1 to a subset of V_2.

Every subset of V_1 in the graph G of Figure 5.4 has negative or zero deficiency; that is, no subset of V_1 has positive deficiency. Hence,

(1) V_1 is nondeficient.

We have already observed that

(2) V_1 is matched to a subset of V_2.

We shall soon see that observations (1) and (2) are synonymous.

Returning to the Dancing Problem, we can now see that we have represented the situation by a bipartite graph G. If we denote by V_1 the vertices corresponding to the women, and by V_2 those vertices corresponding to the men, then we are asking conditions under which V_1 can be matched to a subset of V_2. We now give such conditions.

Theorem 5.4

Let G be a bipartite graph whose vertex set is partitioned into sets V_1 and V_2 so that every edge of G joins a vertex of V_1 with a vertex of V_2. Then V_1 can be matched to a subset of V_2 if and only if V_1 is nondeficient.

The proof of Theorem 5.4 is relatively complicated, and we omit it. It may be found in reference [1], however.

Applying Theorem 5.4 to the Dancing Problem, we can determine whether compatible dancing partners exist for all the women by computing the total number of compatible dancing partners for each subset of women. If the number of compatible dancing partners for each such subset is at least as large as the number of women in the subset, then compatible dancing partners can be found for all women for a single dance.

Example 5.1

Horace Barney is having a few friends over for drinks, namely Al, Bob, Chuck, Dave, Ed, and Frank. Horace has prepared some drinks prior to their arrival, but only one drink of a kind. The drinks are a daiquiri (d), a grasshopper (g), a Mai Tai (m), a Singapore sling (s), a Tom Collins (t), a whiskey sour (w), and a Barney special (b). Horace asks his friends which of these drinks they like, and their answers are: Al (g, s); Bob (g, t, w); Chuck (s, t, w); Dave (d, g, m, s); Ed (g, s, t, w); Frank (g, s, t). Is it possible for all of Horace's friends to have a drink they like?

This situation is represented by the graph of Figure 5.5, where the vertices corresponding to Horace's friends

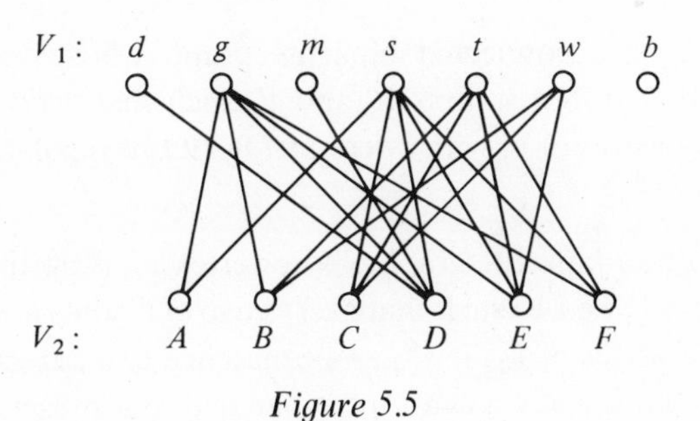

Figure 5.5

(the set V_2) are denoted by the first letters of their names. The problem is whether we can match V_2 to a subset of V_1. By Theorem 5.4, this is possible if and only if V_2 is nondeficient.

Let $U_2 = \{A, B, C, E, F\}$. Then $U_2^* = \{g, s, t, w\}$ and

$$|U_2| - |U_2^*| = 5 - 4 = 1,$$

so U_2 has positive deficiency. Hence, V_2 fails to be non-deficient, which implies that not all of Horace's friends will have drinks they like.

Problem Set 5.2

17. Draw all the (non-isomorphic) bipartite graphs of order 4.

18. Show that the maximum size of a bipartite graph of order 10 is 25.

*19. Determine the maximum size of a bipartite graph of order $p\,(\geq 2)$.

*20. Prove that every tree is a bipartite graph.

*21. Let G be a connected bipartite graph whose vertex set is partitioned into subsets V_1 and V_2 such that every edge of G joins a vertex of V_1 with a vertex of V_2. If G is regular, show that $|V_1| = |V_2|$.

22. Let G be a bipartite graph whose vertex set is partitioned into subsets V_1 and V_2 such that every edge of G joins a vertex of V_1 with a vertex of V_2. If V_1 can be matched to a subset of V_2 and $U_1 \subseteq V_1$, $U_1 \neq \varnothing$, can U_1 be matched to a subset of V_2?

*23. Let G be a bipartite graph whose vertex set is partitioned into subsets V_1 and V_2 such that every edge of G joins a vertex of V_1 with a vertex of V_2. Suppose V_1 can be matched to a subset of V_2. How can this property be described in terms of functions?

24. There are five women at a party [Alice (A), Barb (B), Connie (C), Debbie (D), and Edith (E)] with six men [Albert (a), Ben (b), Charles (c), David (d), Ed (e), and Frank (f)]. The compatible dancing partnerships are displayed by means of a graph in Figure 5.6. Is it possible to have all women dancing so that each dancing partnership is a compatible one?

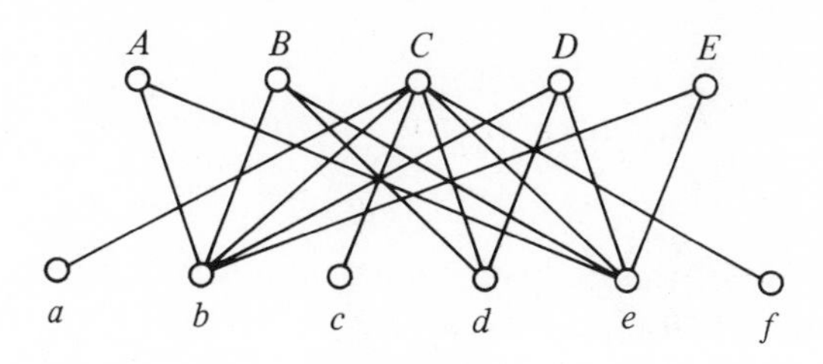

Figure 5.6

25. The following problem is sometimes referred to as ***The Marriage Problem***: Suppose we have a group of men and an equal number of women, and each man is acquainted with a certain number of the women. Under what conditions is it possible for all these men and women to get married, if each marriage is to take place between a couple already acquainted with each other?

26. How many different committees of four members can be formed from a group of five people? Is it possible to have a different chairperson for each committee?

27. A high school has openings for six teachers, with one teacher needed for each of these areas: mathematics, chemistry, physics, biology, psychology, and ecology. In order for a teacher to be hired in any particular area, he or she must have

either majored or minored in that subject. The school receives six applications for these positions, namely: Mr. Arrowsmith (major: physics; minor: chemistry), Mr. Beckman (major: biology; minors: physics, psychology, ecology), Miss Chase (major: chemistry; minors: mathematics, physics), Mrs. Deerfield (majors: chemistry, biology; minors: psychology, ecology), Mr. Evans (major: chemistry; minor: mathematics), Ms. Form (major: mathematics; minor: physics). What is the largest number of applicants the school can hire?

Suggestions for Further Reading

In recent years the study of Ramsey numbers has developed into a theory, namely Ramsey Theory. Articles by Burr [2] and Harary [3] and a chapter in the book by Behzad, Chartrand and Lesniak-Foster [1] show the types of problems which have been investigated.

As indicated earlier, the subject of matchings is not a particularly easy one. More information on the subject can be obtained by reading Ore [4] and Wilson [5].

[1] M. Behzad, G. Chartrand, and L. Lesniak-Foster, *Graphs & Digraphs*. Prindle, Weber & Schmidt, Boston. To be published.

[2] S. A. Burr, "Generalized Ramsey theory for graphs—a survey," *Graphs and Combinatorics* (R. A. Bari and F. Harary, eds.). Springer-Verlag, Berlin (1974), pp. 52–75.

[3] F. Harary, "Recent results on generalized Ramsey theory for graphs," *Graph Theory and Applications* (Y. Alavi, D. R. Lick, and A. T. White, eds.). Springer-Verlag (1972), pp. 125–138.

[4] O. Ore, *Graphs and Their Uses*. Random House, New York (1963).

[5] R. J. Wilson, *Introduction to Graph Theory*. Academic Press, New York (1972).

Chapter 6

Games and Puzzles

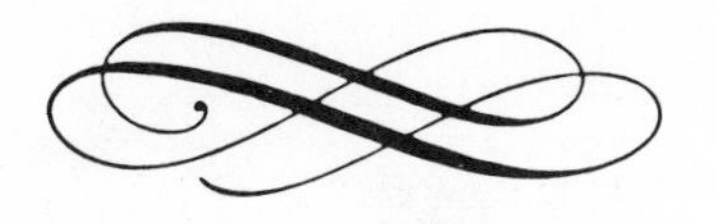

Thus far, the problems we have encountered have been serious problems, for the most part, and several have potentially important implications. However, part of the entertaining aspect of graph theory lies in its usefulness for analyzing certain kinds of games and puzzles. In this chapter we look into a few of the less serious applications of graphs.

6.1 The Problem of the Four Multicolored Cubes: A Solution to "Instant Insanity"

 **The "Instant Insanity" Problem** 

THE PUZZLE "INSTANT INSANITY" (which is a trade name used by the Parker Brothers Game Company) consists of four cubes, and each cube's six faces are colored from four given colors:

red, white, green, and blue. The object of the puzzle is to stack these cubes on top of one another in the form of a $1 \times 1 \times 4$ rectangular prism so that each of the four colors appears on each of the prism's four sides.

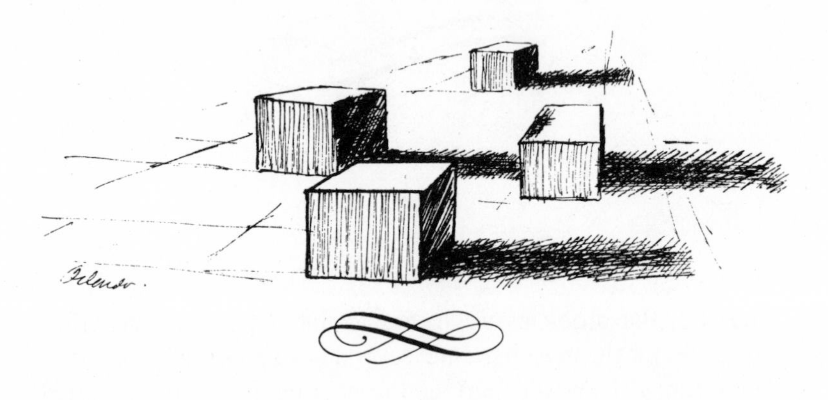

It is convenient first to describe how we will represent a cube. Each face f has an "opposite" face f'. Hence, the cube in Figure 6.1(a) may be represented as in Figure 6.1(b). Thus, in the cube of Figure 6.1(a), we can assume α represents the front of the cube and α' the back; β and β' represent the right and left faces; and γ and γ' represent the top and bottom (see Figure 6.1(c)).

There are three ways to position a cube (the first cube) at the bottom of the stack with respect to which pair of faces appears on the top and bottom of the cube. Faces γ and γ' might be in the top and bottom positions (as in Figure 6.1); that is, they might be the "buried" faces. (Hence the four remaining faces are the visible ones involved in the solution of the problem.) On the other hand, faces β and β' might be buried, or α and α' might be buried.

Suppose we have selected a position for the first cube. There are now 24 possible positions for the second cube. For example, if face α appears as the front face of the second cube, any of the positions shown in Figure 6.2 might occur. Hence, there are four possible arrangements of this cube for each face selected as the front face.

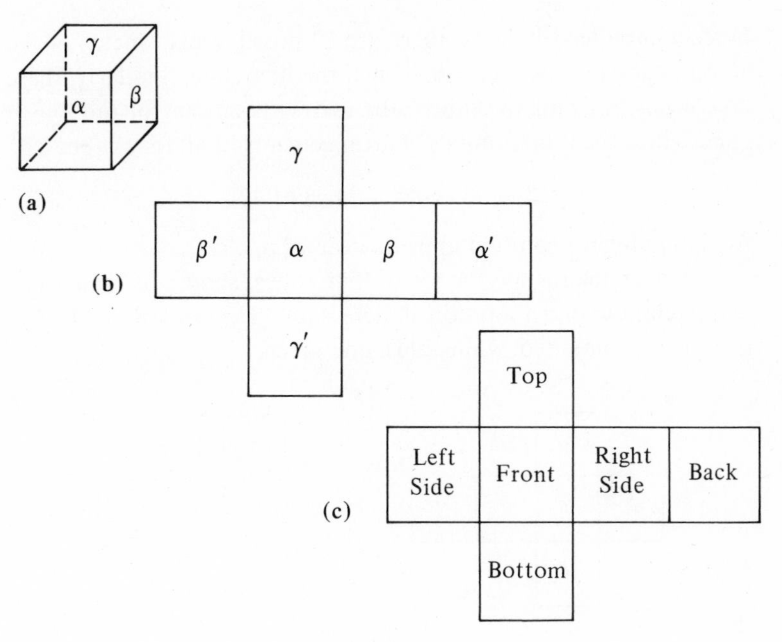

Figure 6.1

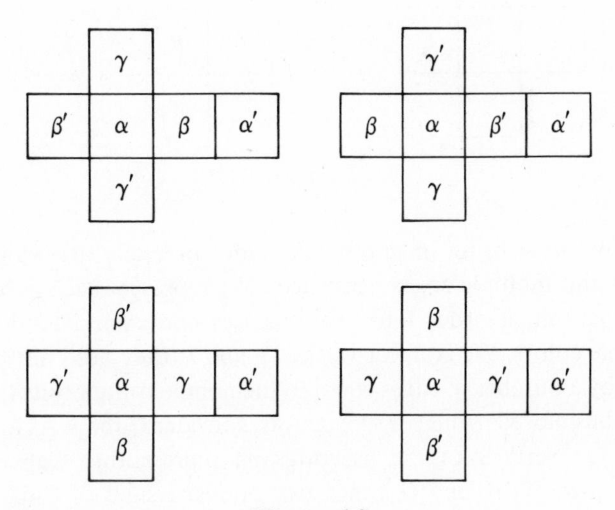

Figure 6.2

Since a cube has six faces, there are 24 possible placements of the second cube once we have stationed the first cube. Similarly, there are 24 positions for the third cube and 24 positions for the fourth cube. Thus, the total number of arrangements of all four cubes is

$$3 \times 24 \times 24 \times 24 = 41{,}472.$$

A trial-and-error solution appears nearly hopeless.

Let us take a specific set of four cubes, shown in Figure 6.3, and attempt to find a solution, if there is one. The symbols R, W, B, G denote the colors red, white, blue, and green.

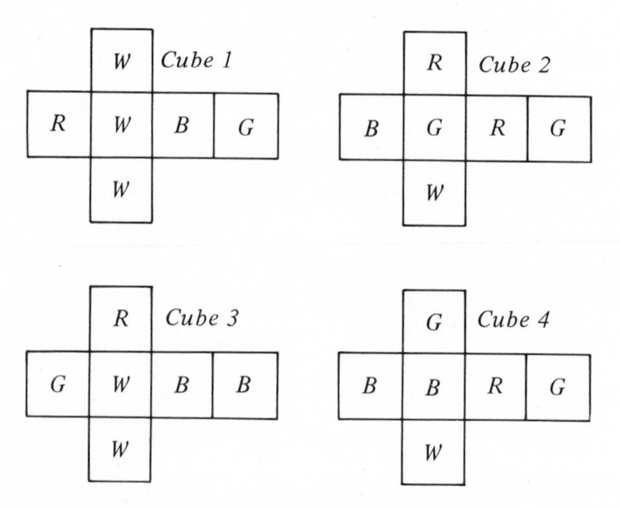

Figure 6.3

We now bring in graphs, or, more precisely, pseudographs (loops and multiple edges permitted). We represent each cube by a pseudograph of order four. The vertices correspond to the four possible colors. We connect vertices u and v (they need not be distinct) by a number of edges equal to the number of opposite faces in the cube colored u and v. For example, consider Cube 1. As for each cube, the vertices of the pseudograph representing Cube 1 are denoted R, W, B, and G. Since two opposite sides of Cube 1 are colored red and blue, R is adjacent to B in this pseudograph. Since

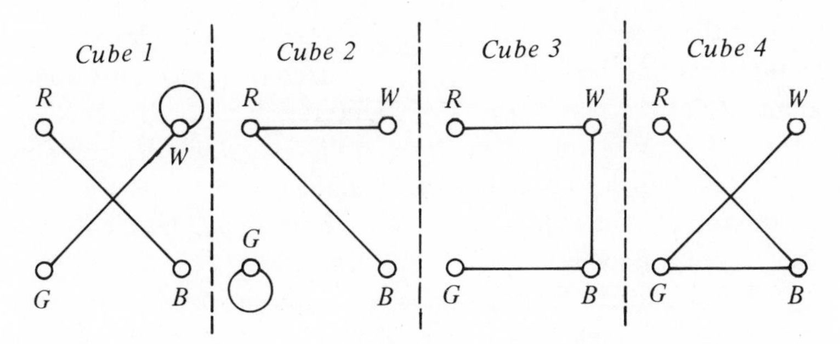

Figure 6.4

the front and back are colored white and green, W is adjacent to G. Finally, since the top and bottom of Cube 1 are both colored white, we add a loop at W. The four pseudographs shown in Figure 6.4 correspond to the four cubes given in Figure 6.3.

We now superimpose the four pseudographs of Figure 6.4, obtaining a pseudograph with 4 vertices and 12 edges. This new pseudograph is shown in Figure 6.5. Each edge is labeled by the number of the cube associated with it.

Let us assume there is a solution to the problem; that is, assume we can stack the four cubes in such a way that all four colors appear on all four sides of the prism. Consider first the front of this prism

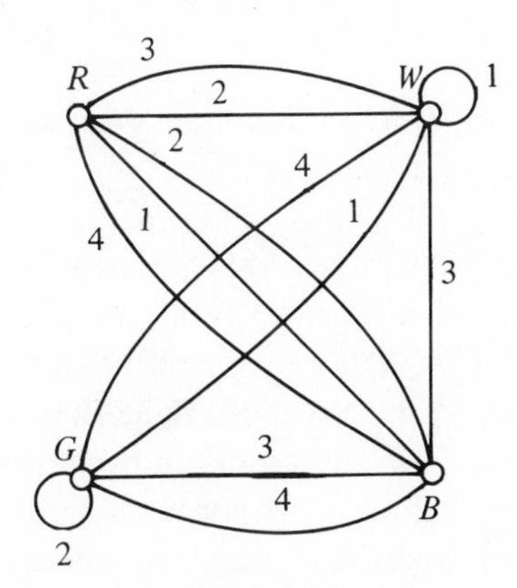

Figure 6.5

(and, at the same time, the back of the prism). Each color appears on this front side. If a solution exists, we claim there is a sub-pseudograph H_1 of the pseudograph of Figure 6.5 which represents the front and back of this prism. The sub-pseudograph H_1 contains four edges, with each edge labeled by a different number, 1, 2, 3, or 4. For example, the edge labeled 1 joins the front and back colors of the first cube. Each vertex of H_1 has degree two (a loop at a vertex contributes a degree of two), since a color appears once on the front and once on the back. Hence H_1 is a 2-regular pseudograph of order 4. Similarly, the left and right sides of the prism give rise to a 2-regular pseudograph H_2 of order 4 which is edge disjoint from H_1 (i.e., H_1 and H_2 have no edges in common). Hence, the problem is to determine two such pseudographs H_1 and H_2. In our particular case, there are such pseudographs, shown in Figure 6.6.

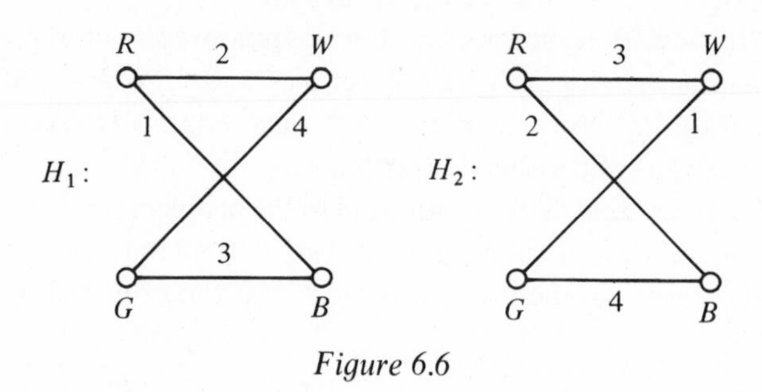

Figure 6.6

With the aid of H_1 and H_2 of Figure 6.6, we now show how to stack the cubes. We start with the first cube, and place red in the front, say, and blue in the back. Place green on the left, say, and white on the right. (Of course, we could just as well begin by switching left and right, or front and back.) It is convenient to consider the third cube next. Here we place blue in front and green in back and we put white on the left and red on the right. Returning to the second cube, we place white in front and red in back. Red appears on the left, and blue on the right. Finally, for the fourth cube, we set green in front and white in back, while blue is on the left and green on the right.

The solution to "Instant Insanity" suggested here still involves a certain amount of work. With each set of four cubes, we construct a pseudograph of order 4 and size 12 in the manner described above. A solution to the puzzle involves locating two edge-disjoint, 2-regular sub-pseudographs. Once these two sub-pseudographs are found, care must be taken in stacking the cubes. However, this remains considerably simpler than trying all 41,472 arrangements of the four cubes.

Problem Set 6.1

*1. Show that the pseudograph of Figure 6.5 does not contain three edge-disjoint, 2-regular sub-pseudographs of order 4.

2. In the description at the end of this section of how to stack the four cubes, why was the third cube considered before the second cube?

3. Solve "Instant Insanity" with the four cubes pictured in Figure 6.7.

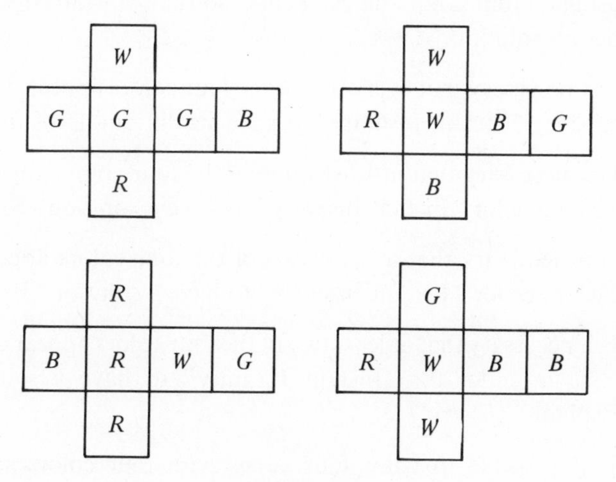

Figure 6.7

4. Solve "Instant Insanity" with the four cubes pictured in Figure 6.8. Show that there are three solutions.

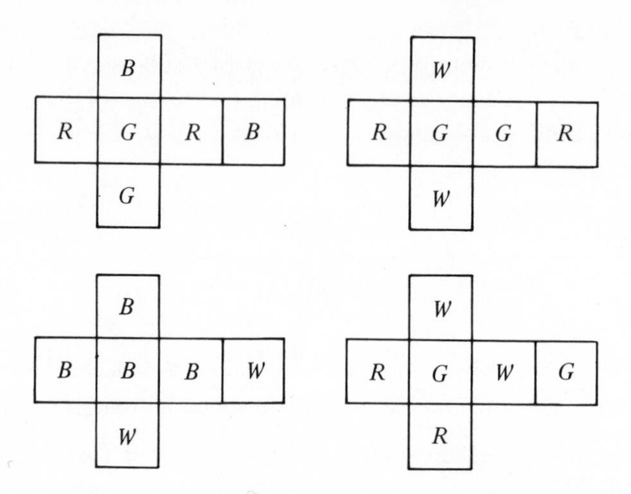

Figure 6.8

5. Show that it is possible to color four cubes with four colors, using all four colors on each cube, so that "Instant Insanity" has no solution.

6. Is it necessary that every one of the four colors appear on each cube for "Instant Insanity" to have a solution? Explain.

7. Is it necessary that at least three of the four colors appear on each cube for "Instant Insanity" to have a solution? Explain.

8. Is it necessary that at least two of the four colors appear on each cube for "Instant Insanity" to have a solution? Explain.

9. Is it necessary that at least two of the four colors appear on any cube in order for "Instant Insanity" to have a solution? Explain.

*10. Is it possible to color four cubes with four colors so that "Instant Insanity" has a unique solution? Explain.

11. It was mentioned that there are 41,472 possible arrangements of all four cubes. Exercise 5 implies that for certain sets of cubes, all 41,472 arrangements may fail to yield a solution. What is the *maximum* number of solutions possible from these 41,472 arrangements?

*12. Is it possible to have an "Instant Insanity" game involving more than four cubes? If so, what restrictions might you place on the cubes to make the game solvable? What restrictions would make the game more interesting?

6.2
The Knight's Tour

Consider the standard 8-by-8 chessboard, with squares colored alternately white and black. Suppose we place a knight on one of these 64 squares. According to the rules of chess, a knight moves by proceeding two squares vertically or horizontally from its starting square, followed by moving one square in a perpendicular direction. We can now state our game:

The Knight's Tour Puzzle

FOLLOWING THE RULES OF CHESS, is it possible for a knight to tour the chessboard, visiting every square once and only once, and return to its initial square?

This problem dates back to the time of Euler, although in recent years it has been merchandized as a game. The Knight's Tour Puzzle is equivalent to determining whether a certain graph G is hamiltonian, where the vertices u_i of G correspond to the squares S_i of the chessboard, and u_i and u_j are adjacent if and only if it is possible for a knight to proceed from S_i to S_j in a single move. The graph G has 64 vertices and 168 edges, and actually contains several hamiltonian cycles, one of which is shown in Figure 6.9. Thus we can show an affirmative solution to the Knight's Tour Puzzle.

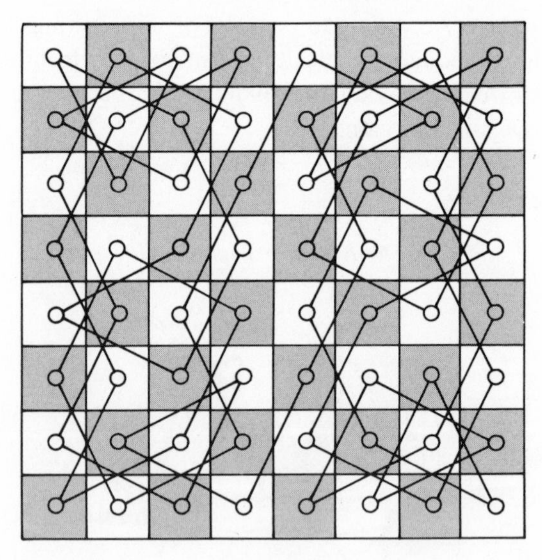

Figure 6.9

Problem Set 6.2

13. In the graph G of the knight's moves on the chessboard, determine the length (number of edges) of the shortest path from the vertex of a corner square to the vertex of the diagonally opposite corner square.
14. Investigate the Knight's Tour Puzzle for a 4-by-4 chessboard.
15. Investigate the Knight's Tour Puzzle for a 4-by-5 chessboard.
16. Investigate the Knight's Tour Puzzle for an n-by-n chessboard, where n is odd.

*17. For which other chess pieces does a corresponding tour exist?

6.3
The Tower of Hanoi

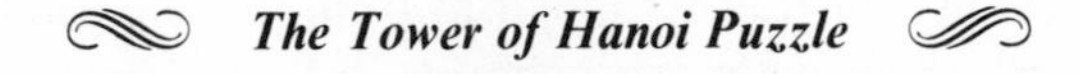

The Tower of Hanoi Puzzle

SUPPOSE WE FASTEN THREE PEGS vertically on a stand, and suppose we have eight circular disks, each containing a hole in the middle through which we can pass a peg. Assume each disk has a different radius and, initially, we place all of them on a single peg so that the radii increase as we proceed from the top disk to the bottom (see Figure 6.10). This stack of disks is called the Tower of Hanoi.

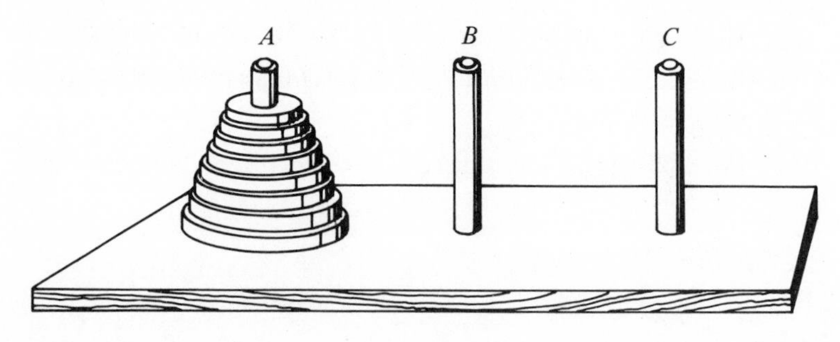

Figure 6.10

The puzzle, also known as the Tower of Hanoi (introduced by M. Claus in 1883), is to transfer the disks from the one peg to another in such a way that we move only one disk at a time, and at any stage in the transfer, no disk lies on a smaller disk. Game manufacturers have produced this puzzle under such names as the "Tower of Trouble" and the "Tree Puzzle" (not referring to the graph-theoretic tree).

The procedure for solving this puzzle is as simple to describe for n disks as for 8; thus, we consider a solution for the general case. Suppose all n disks begin on peg A, and assume we require x moves to transfer the top $n - 1$ disks to peg B, leaving the largest disk on peg A and leaving peg C empty. We may now move the largest disk to peg C. As before, we need x moves to transfer the $n - 1$ disks on peg B to peg C, solving the problem. This has taken us $2x + 1$ moves in all. By considering the puzzle for small n, we see that we require $2^n - 1$ moves to solve the puzzle, or, equivalently, there are 2^n stages in all from the initial to the final position. So, for the 8-disk puzzle, it takes us $2^8 - 1 = 255$ moves.

Thus far, we have said nothing about graphs. As we shall see presently, there is a connection between the solution to the Tower of Hanoi puzzle just presented and hamiltonian graphs. To understand this connection, we first introduce a new class of graphs.

We mentioned the graph of the cube in Exercise 19 in Chapter 3. This is the graph obtained, very naturally, by representing the vertices and edges of the cube in the plane. This cube is known as the 3-dimensional cube or 3-cube. The associated graph is also

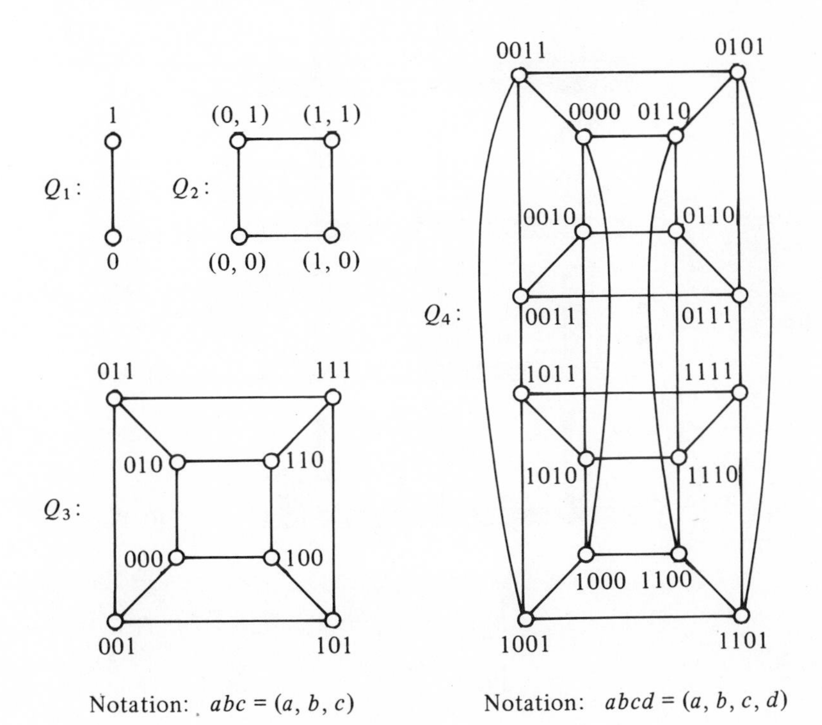

Figure 6.11

called the 3-cube. This idea has been generalized to the *n-cube* Q_n, which is the graph of order 2^n whose vertices are represented by n-tuples (a sequence of n numbers). Each term of the n-tuple has the value 0 or 1, and two vertices of Q_n are adjacent if and only if the corresponding n-tuples differ in only one term. Figure 6.11 shows Q_1, Q_2, Q_3, and Q_4.

To represent a solution of the Tower of Hanoi by an n-cube, we number the disks 1, 2, . . . , n from the smallest disk to the largest

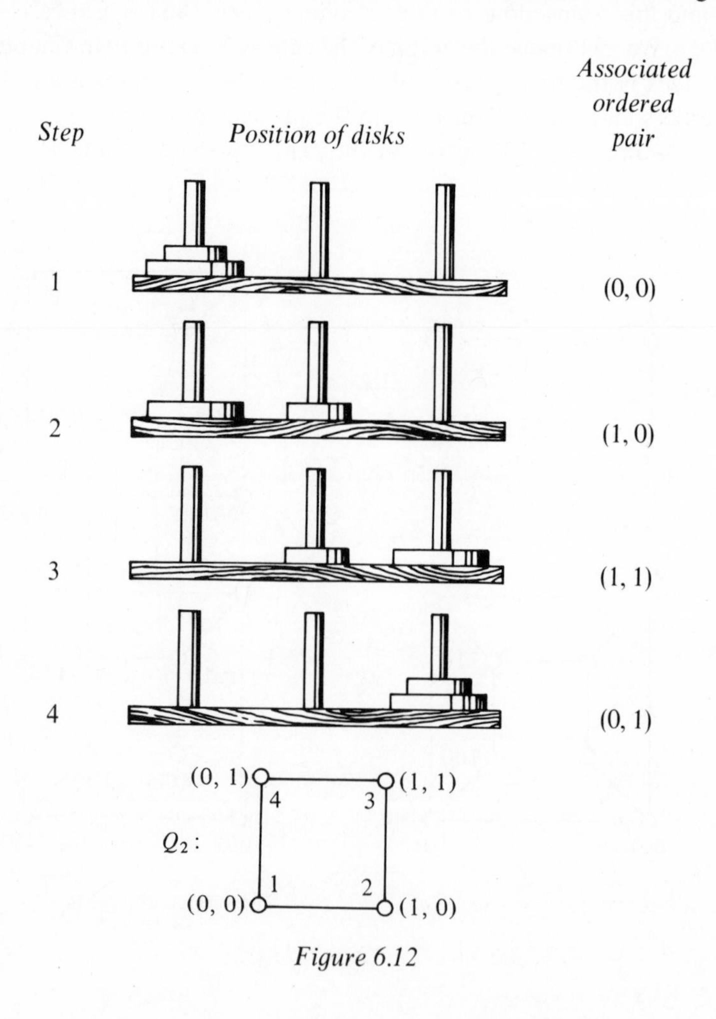

Figure 6.12

disk. With each step in the solution, we associate an n-tuple as follows. Let the jth position of the disks in the solution to the puzzle be represented by $(a_{1j}, a_{2j}, \ldots, a_{nj})$, where a_{ij} is the number of moves, modulo 2 (meaning $1 + 1 = 0$), made by the ith disk. It turns out that the sequence of n-tuples so described, in the order in which they occur in the solution of the puzzle, produces a hamiltonian cycle in Q_n.

Let us illustrate this relationship between hamiltonian graphs and the Tower of Hanoi with only two disks. The steps in the solution, together with the associated pairs and Q_2, are shown in Figure 6.12.

Problem Set 6.3

18. Draw the graph Q_5 and label the vertices with the appropriate 5-tuples.

19. Solve the Tower of Hanoi puzzle for $n = 3$, and determine the corresponding hamiltonian cycle of Q_3.

*20. Prove that Q_n is hamiltonian for $n \geq 2$. (*Hint*: Use mathematical induction.)

6.4
The Three Cannibals and Three Missionaries Problem

The solution to the Tower of Hanoi Puzzle depends on various positions, or "states," in the problem. The solution proceeded from an initial state through a sequence of intermediate states to a final state. We used graphs to solve this problem by associating a vertex with each state. Two vertices were joined by an edge whenever we could proceed from one corresponding state to another in a single "move." Then, in the resulting graph G, we attempted to determine a path from the vertex corresponding to the initial position to the vertex corresponding to the final position. We now consider another problem that falls into this category.

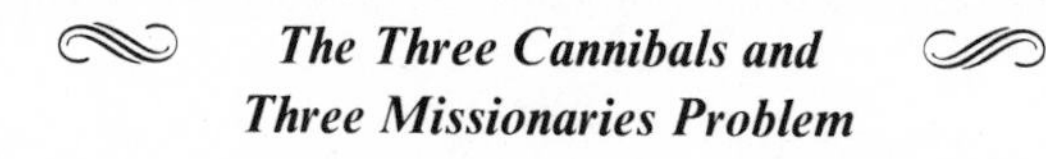

The Three Cannibals and Three Missionaries Problem

THREE CANNIBALS and three missionaries are traveling together, and they arrive at one bank of a river. All three cannibals and all three missionaries want to cross the river to the opposite side. However, the only transportation across the river is a boat which holds at most two people. There is another difficulty: At no time, on neither bank, can the cannibals outnumber the missionaries, for then the missionaries would be in danger. How do they manage to cross the river?

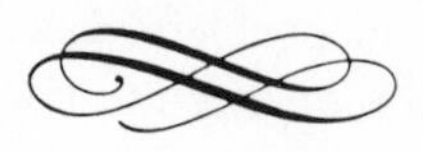

We now present a graphical solution to this problem. By a "position" or "state" here, we mean the situation as it occurs on the first bank. Let c represent the number of cannibals on the first bank and m represent the number of missionaries on the first bank. Hence, the ordered pair (c, m) denotes the state of the system at any one time.

First, we make a few observations. Note that $0 \leq c \leq 3$ and $0 \leq m \leq 3$. Hence, there are 16 possible ordered pairs (c, m). However, some of these states are not permitted under the additional restrictions of the problem. For example, in the cases (0, 1), (0, 2), (1, 2), (2, 1), (3, 1), and (3, 2), the cannibals would outnumber the missionaries on the first bank or on the second bank. The permissible ordered pairs are given in Figure 6.13.

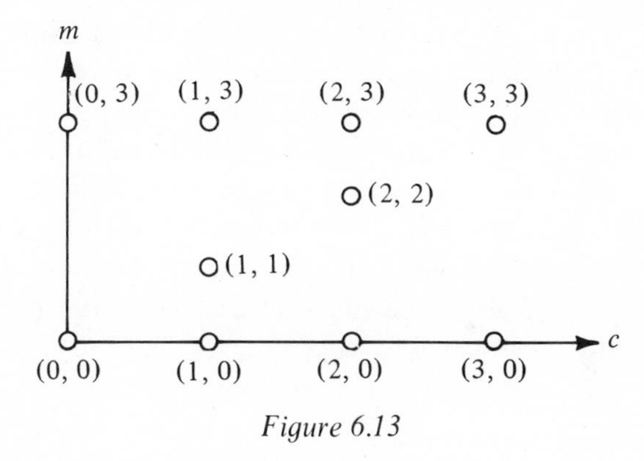

Figure 6.13

We must proceed from the ordered pair (3, 3) to (0, 0), but we must do this according to certain rules. If we are at the state (c_1, m_1) and the boat is at the first bank, then we can proceed to the permissible pair (c_2, m_2) provided $c_2 \leq c_1$, $m_2 \leq m_1$, and

$$(c_1 + m_1) - (c_2 + m_2) \leq 2.$$

If we are at the state (c_1, m_1) and the boat is at the further bank, then we can proceed to the permissible pair (c_2, m_2) provided $c_2 \geq c_1$, $m_2 \geq m_1$, and

$$(c_2 + m_2) - (c_1 + m_1) \leq 2.$$

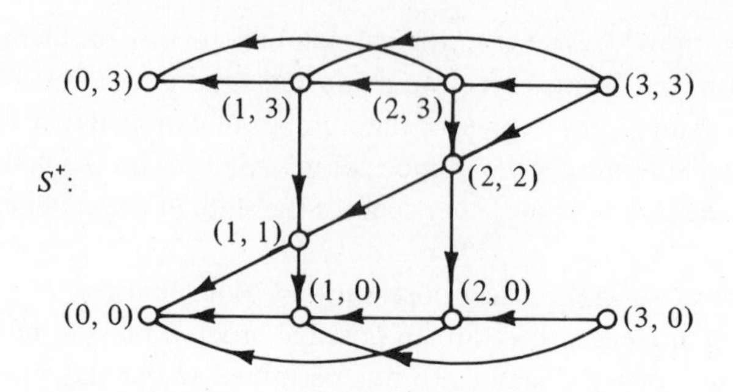

Figure 6.14

For example, if we are at the state (2, 2) and the boat is at the first bank, then the next state could be (1, 1) or (2, 0).

It is convenient to introduce a directed signed graph S. The vertices of S correspond to the ten permissible states. We join a positive edge from (c_1, m_1) to (c_2, m_2) if we can proceed from state (c_1, m_1) to state (c_2, m_2) by moving the boat from the first bank to the second bank. We direct a negative edge from (c_1, m_1) to (c_2, m_2) if we can proceed from state (c_1, m_1) to state (c_2, m_2) by moving the boat from the second bank to the first bank. The directed graph S^+, consisting of the positive edges of S, is shown in Figure 6.14. Observe that there is a positive edge from (c_1, m_1) to (c_2, m_2) if and only if there is a negative edge from (c_2, m_2) to (c_1, m_1).

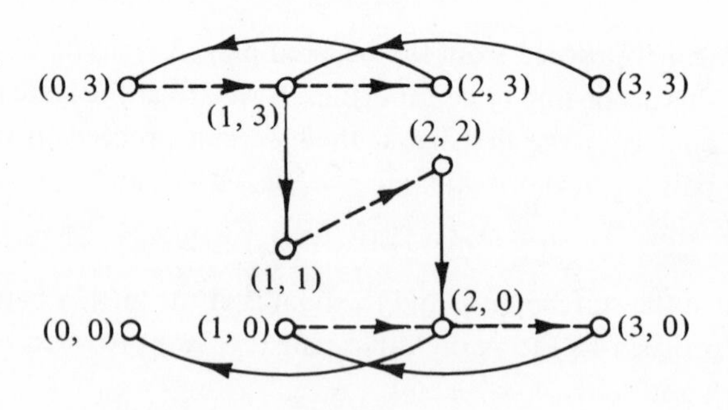

Figure 6.15

The object of the problem is to find a directed trail in S from (3, 3) to (0, 0) which alternates positive and negative edges. One such trail is shown in Figure 6.15. Figure 6.16 presents a diagram of the eleven crossings described by the trail of Figure 6.15, showing the number of cannibals and the number of missionaries on each river bank after each crossing.

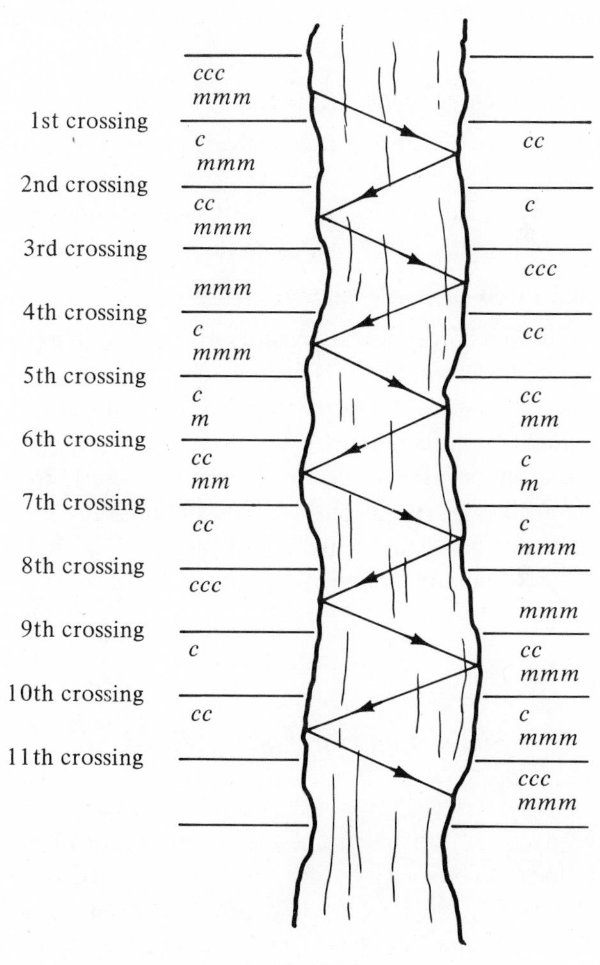

Figure 6.16

Problem Set 6.4

*21. Show that there is no solution of the Three Cannibals and Three Missionaries Problem requiring less than eleven river crossings.

*22. Would it make sense to talk about the Four Cannibals and Four Missionaries Problem? If so, explain the conditions of the problem and attempt to solve it.

23. Solve the ***Three Jealous Husbands Problem***:

    Three wives and their three husbands must get to town in a Corvette, which holds only two people. How might they do this so that no wife is ever left with one or both of the other husbands unless her own husband is also present?

24. Solve the ***Wine Bottle Problem***:

    Three wine bottles have capacities of 8, 5, and 3 liters, respectively. The largest bottle is filled with wine and the other containers are empty. We want to divide the wine into two equal portions using these bottles (which are not graduated) and no others, by pouring successively from one bottle to another. The problem, then, is: How can we obtain 4 liters of wine in the largest bottle and 4 liters in the medium-size bottle, using the fewest possible number of pourings?

Suggestions for Further Reading

We have given only a few examples of the multitude of games and puzzles which can be solved by graphs. Of the several descriptions of "Instant Insanity," the reader might wish to consult Busacker and Saaty [2, p. 153] and VanDeventer [6]. An excellent description of the Knight's Tour Problem is presented in Berge [1, p. 109], and

Crowe [3] has written an article on the Tower of Hanoi. Problems similar to the Three Cannibals and Three Missionaries Problem are described in Ore [5, p. 30]. For an enormous variety of puzzles involving graphs, however, the reader is recommended to consult the interesting pamphlet written by Fujii [4].

[1] C. Berge, *The Theory of Graphs.* Wiley, New York (1962).

[2] R. G. Busacker and T. L. Saaty, *Finite Graphs and Networks.* McGraw-Hill, New York (1965).

[3] D. W. Crowe, "The *n*-dimensional cube and the tower of Hanoi." *American Mathematical Monthly*, **63** (1956), pp. 29–30.

[4] J. N. Fujii, *Puzzles and Graphs.* National Council of Teachers of Mathematics, Washington D.C. (1966).

[5] O. Ore, *Graphs and Their Uses.* Random House, New York (1963).

[6] J. VanDeventer, "Graph theory and 'Instant Insanity'." *The Many Facets of Graph Theory* (G. Chartrand and S. F. Kapoor, eds.). Springer-Verlag, Berlin (1969).

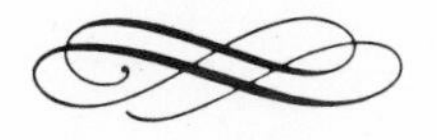

Chapter 7

Digraphs and Mathematical Models

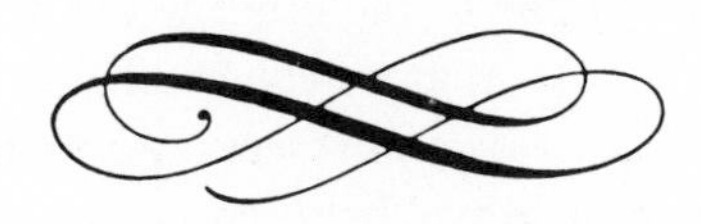

As we mentioned in Chapter 1, there are several real-life situations and problems where digraphs are more appropriate mathematical models than graphs. In this chapter we illustrate a few such situations and problems. We include a discussion of a much studied class of digraphs: the tournaments.

7.1
A Traffic System Problem: An Introduction to Orientable Graphs

Suppose we wish to design a traffic system for a small college town in the East. Since the town is small and traffic is ordinarily light, a complex traffic system is not needed; so we agree to make all streets two-way. Certain minimum requirements are necessary; for example, we want to be able to reach any intersection from any other intersection. Also, the street crews make any necessary road repairs during the week. If a crew is repairing a street between two consecutive intersections, thereby blocking off the street, it would be convenient if a motorist could still reach any intersection while the repairs are in progress.

We can easily represent this situation by a graph G. The vertex set of G corresponds to the street intersections, and two vertices are joined by an edge if, in the corresponding two intersections, it is possible to travel from one to the other without passing through a third intersection. The fact that we can reach any intersection in the traffic system from any other intersection implies that G must be connected. If we want to travel between any two intersections even when a street between consecutive intersections is blocked off, then the graph G can have no bridges.

Let us now suppose that on football Saturdays, traffic becomes so heavy we need a different traffic pattern. We decide to convert all streets into one-way streets. Of course, we would like to make this change in such a way that afterwards, using one-way streets, it is still possible to travel (legally) from any intersection to any other intersection. This leads us to our problem.

The Traffic System Problem

A Traffic System Problem: An Introduction to Orientable Graphs

UNDER WHAT CONDITIONS can a traffic system with all two-way streets be changed to all one-way streets so that in the resulting system, it is possible to travel from any intersection to any other intersection?

In order to solve this problem, we first present a few definitions. Recall that a path in a directed graph is an alternating sequence $v_1, a_1, v_2, a_2, v_3, \ldots, v_{n-1}, a_{n-1}, v_n$ of distinct vertices and arcs such that arc $a_i = (v_i, v_{i+1})$ for $i = 1, 2, \ldots, n-1$. A digraph D is called *strongly connected* if for every two distinct vertices u and v of D, there exists a u-v path in D *and* a v-u path in D.

A connected graph G is called *orientable* if it is possible to assign a direction to each edge of G (thus changing edges to arcs) to produce a strongly connected digraph D.

We now present a solution to the traffic problem. The solution may be somewhat unexpected, for it states that if a traffic system is suitable for travel during street repairs on weekdays, then it is also suitable for one-way street conversion on football Saturdays, and conversely.

Theorem 7.1
Solution to the Traffic System Problem

A connected graph G is orientable if and only if G contains no bridges.

Proof

First we prove that if G .s orientable, then G cannot contain a bridge. We prove this by contradiction. Thus, let G be an orientable graph and assume, to the contrary, that G contains a bridge, say $e = uv$. Since G is orientable, we can assign a direction to each edge of G and obtain a strongly connected digraph D. Hence, D contains both a u-v path and a v-u path. Since uv is a bridge, it is an edge of G, so the digraph D contains exactly one of the arcs (u, v) and (v, u). Without loss of generality, we may assume that D contains (u, v). Then, of course, D contains a u-v path, namely u, v. However, D cannot contain a v-u path, since every path from v to u must contain the arc (v, u) because uv is a bridge of G. Therefore G is not orientable, giving us a contradiction. Hence G cannot contain bridges.

Next we verify the converse statement. Assume that G is a connected graph containing no bridges. We prove that G is orientable. Since G contains no bridges, it follows by Theorem 2.5 that every edge of G lies on a cycle. Let $C: v_1, v_2, \ldots, v_n, v_1$ be a cycle of G. We assign the edge $v_n v_1$ the direction (v_n, v_1), and we assign the edge $v_i v_{i+1}$ the direction (v_i, v_{i+1}) for $i = 1, 2, \ldots, n - 1$. If there exist adjacent, nonconsecutive vertices of C, then the edges between them are directed arbitrarily. If every vertex of G belongs to C, then it is obvious that G is orientable.

Assume there exist vertices of G not belonging to C. Since G is connected, there exists a vertex w_1, not on C, such that $w_1 v_j$ is an edge of G, where $1 \leq j \leq n$. Since $w_1 v_j$ is not a bridge of G, there exists a cycle $C_1 : w_1, v_j = w_2, w_3, \ldots, w_m, w_1$ in G. We assign $w_m w_1$ the direction (w_m, w_1) and we assign $w_1 w_2$ the direction (w_1, w_2). For each $i = 2, 3, \ldots, m - 1$ for which $w_i w_{i+1}$ has not already been given a direction, we assign the direction (w_i, w_{i+1}). If an edge joining two vertices of C_1 (or one vertex of C and one vertex of C_1) has not yet received a direction, then a direction can be selected arbitrarily. (See Figure 7.1 for one possibility.)

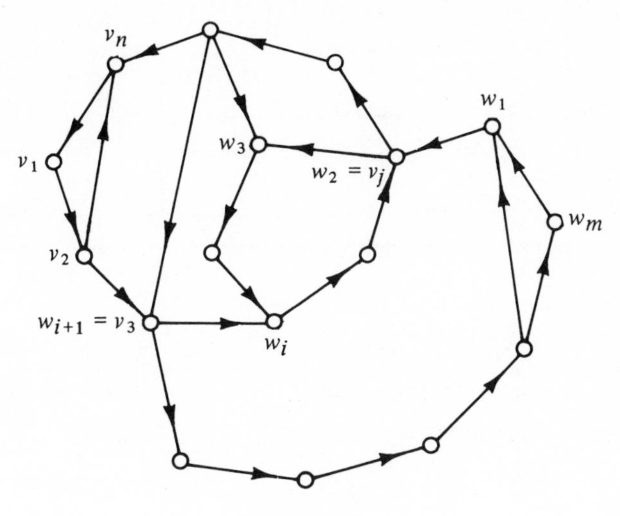

Figure 7.1

The digraph D_1 constructed in this manner is necessarily strongly connected. If D_1 contains all vertices of G, we are finished. If not, we continue this procedure until all vertices of G have been included in a strongly connected digraph D, which completes the proof. ▮

Example 7.1

Figure 7.2 shows the street system of a small town. A graph G which models this traffic system is also given in Figure 7.2. Since G does not contain a bridge, by Theorem 7.1, G is orientable. Thus, we may assign a direction to each edge of G and obtain a strongly connected digraph. The digraph D of Figure 7.2 is strongly connected.

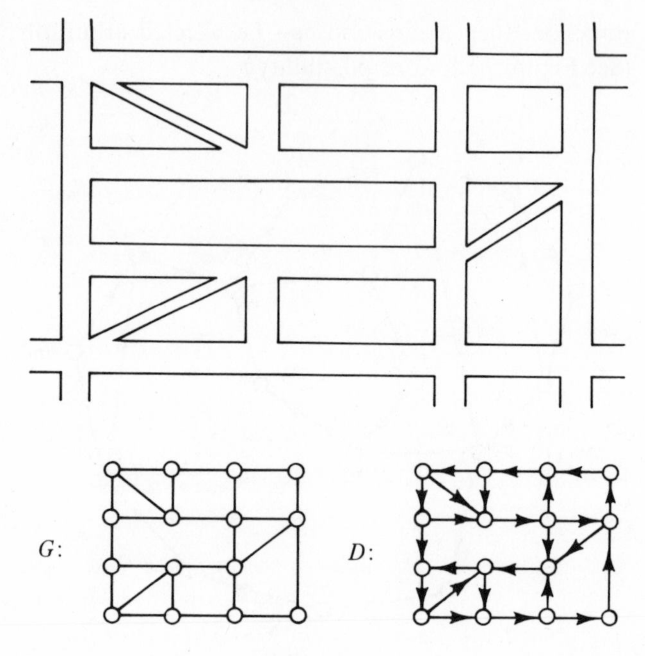

Figure 7.2

Problem Set 7.1

1. Explain why the proof of Theorem 7.1 is particularly simple if the graph G is hamiltonian.

2. Show that the graph G of Figure 7.2 is hamiltonian. Is there a relationship between the digraph D of Figure 7.2 and your answer to Exercise 1?

3. Show that the graph G of Figure 7.3 is orientable. Assign appropriate directions to the edges of G to obtain a strongly connected digraph D.

*4. Let G be an orientable graph, and assign directions to the edges of G to produce a strongly connected digraph D. Suppose

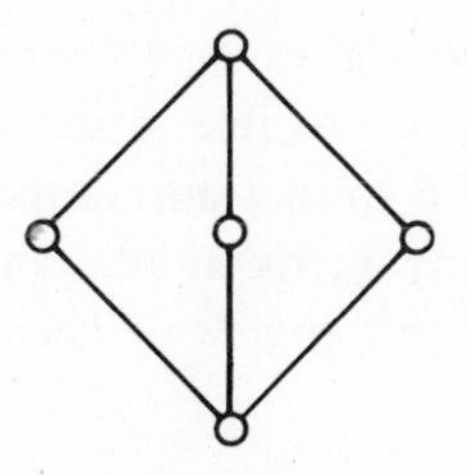

Figure 7.3

that for each edge e of G, we assign the direction opposite to that used in producing D. Let D' denote the digraph obtained in this way. Is D' strongly connected? Prove your answer.

5. Can a graph with cut-vertices be orientable? Explain.
6. Can an orientable graph G of order at least 2 possess the property that no matter how the edges of G are directed, a strongly connected digraph D always results? Explain.
7. Figure 7.4 shows the street system of a town. Investigate this system in the same way as in Example 7.1.

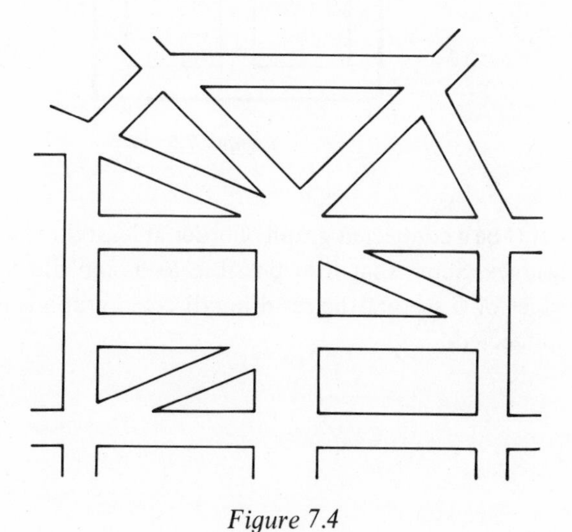

Figure 7.4

8. Figure 7.5 shows the street system of a town where all streets are two-way. Is it possible to repair the road between intersections A and B so that during the repair work it is possible to drive from any intersection to any other intersection? How do you explain this in view of Theorem 7.1?

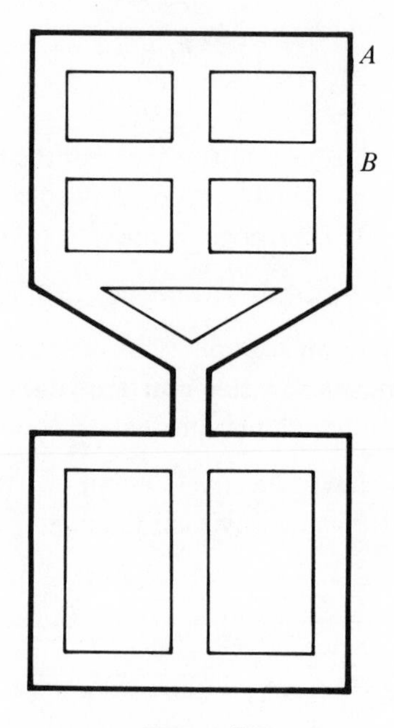

Figure 7.5

9. Let G be a connected graph of order at least three which has no bridges. Show that it is possible to assign directions to the edges of G so that the resulting directed graph is *not* strongly connected.

7.2
Tournaments

In the preceding section we saw how directed graphs may serve as models for traffic control problems. As you might expect, there are many other problems for which directed graphs are useful models. Indeed, there is one particular class of directed graphs that we can use to represent a large variety of situations. These are called "tournaments," the name originating because they represent round robin tournaments. Before investigating some problems relevant to tournaments, we present some pertinent definitions.

A *tournament* T is a directed graph with the property that for every two distinct vertices u and v, exactly one of (u, v) and (v, u) is an arc of T. That is, a tournament is a digraph which results from assigning directions to the edges of a complete graph.

There is only one tournament of order one and only one tournament of order two (i.e., every two tournaments of order two are isomorphic). There are two non-isomorphic tournaments of order three; they are shown in Figure 7.6. One way of seeing that tournaments T_1 and T_2 of Figure 7.6 are different is by observing that T_2 contains a vertex, namely u_2, which is not adjacent from any other vertex, while T_1 contains no such vertex. It is helpful to consider some definitions involving degrees in directed graphs.

Let D be a directed graph and let v be a vertex of D. The number of vertices of D adjacent *from* v is called the *outdegree* of v, denoted

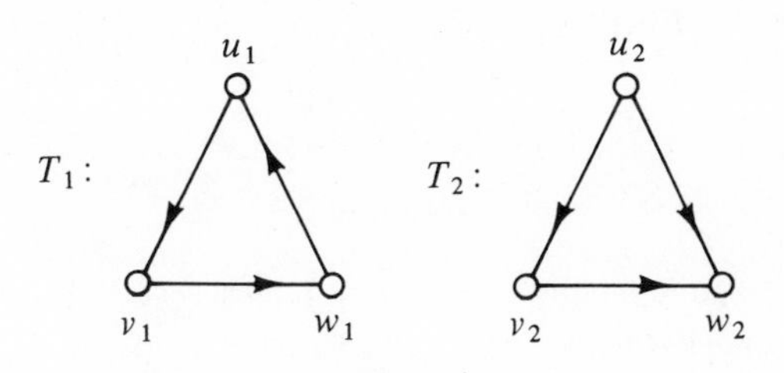

Figure 7.6

by od v; the number of vertices of D adjacent *to* v is called the *indegree* of v, denoted by id v. In the tournament T_1 of Figure 7.6, every vertex has the same indegree and the same outdegree, namely 1. In T_2, the vertices have the following indegrees and outdegrees:

$$\text{od } u_2 = \text{id } w_2 = 2, \qquad \text{od } v_2 = \text{id } v_2 = 1, \qquad \text{od } w_2 = \text{id } u_2 = 0.$$

A *transmitter* is a vertex having positive outdegree and zero indegree; a *receiver* is a vertex having positive indegree and zero outdegree. Thus, in the tournament T_2 of Figure 7.6, vertex u_2 is a transmitter and w_2 is a receiver; tournament T_1 has no transmitters or receivers.

Suppose T is a tournament of order $p \geq 2$, and let v be any vertex of T. Then by $T - v$ we mean the directed graph obtained from T by removing v and all arcs incident with v. Necessarily, any two vertices of $T - v$ are joined by exactly one arc, since these two vertices are joined by one arc in T. Hence, if T is a tournament, $T - v$ is also a tournament. This very elementary property of tournaments turns out to be quite useful.

By a round robin tournament we mean a competition among a collection of teams such that every two teams play each other exactly once and no ties are permitted. We can represent every round robin tournament by a tournament T where the vertices of T correspond to the individual teams and where (u, v) is an arc of T if and only if the team corresponding to u defeats the team corresponding to v.

In order to present our first theorem on tournaments, we need some additional definitions. The *length* of a path P in a digraph is the number of arcs in P. If u and v are two vertices of a directed graph, then the *distance $d(u, v)$ from u to v* is the length of a shortest u-v path.

Theorem 7.2

Let T be a tournament, and let v be any vertex of T having maximum outdegree. Then the distance from v to any other vertex of T is one or two.

Proof

Let od $v = n$, and suppose the vertices v is adjacent to are $v_1, v_2, \ldots, v_n$. If the order of T is p, then each of the remaining $p - n - 1$ vertices $u_1, u_2, \ldots, u_{p-n-1}$ is adjacent to v, since T is a tournament (see Figure 7.7).

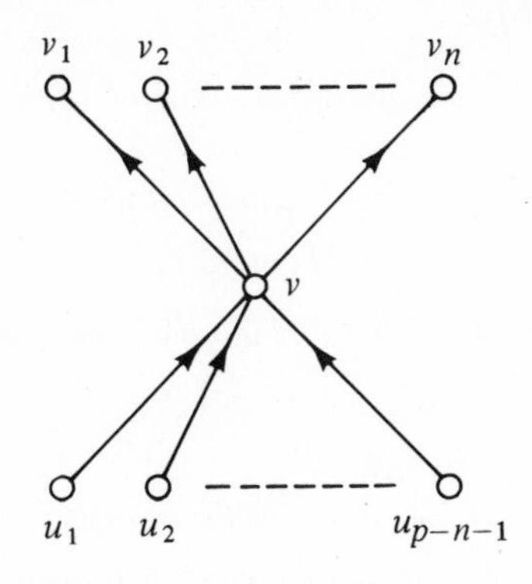

Figure 7.7

For each vertex v_i, where $1 \leq i \leq n$, we have $d(v, v_i) = 1$. We now show that for each vertex u_j, where $1 \leq j \leq p - n - 1$, the distance $d(v, u_j) = 2$. Once this is shown, the proof will be complete.

If each u_j is adjacent from some v_i, then it is obvious that $d(v, u_j) = 2$. Suppose there is a vertex u_k, where $1 \leq k \leq p - n - 1$, such that no vertex v_i, $1 \leq i \leq n$, is adjacent to u_k. Then u_k is adjacent to all the vertices $v_1, v_2, \ldots, v_n$ and to the vertex v as well. However, then od $u_k \geq n + 1$, which is impossible because v has the largest outdegree and od $v = n$. Hence each u_j is adjacent from some v_i. ▮

Theorem 7.2 has the following interpretation in round robin tournaments. Suppose several sports teams participate in a round robin tournament (i.e., every two teams play each other and no ties are permitted). Then the winner w (any team with the most victories —there may be more than one winner) has been defeated only by teams which themselves have lost to teams defeated by w.

Another interesting property of tournaments involves hamiltonian paths. A *hamiltonian path* in a digraph D is a path containing all the vertices of D.

Theorem 7.3

Every tournament contains a hamiltonian path.

Proof

The proof is by induction on the number of vertices p in the tournament. We can actually check out the tournaments with 1, 2, 3, or 4 vertices and verify that each contains a hamiltonian path. So we assume that each tournament of order $n, n \geq 4$, contains a hamiltonian path, and we consider a tournament T of order $n + 1$. We show that T contains a hamiltonian path.

Let v be a vertex of T. As we noted earlier, $T - v$ is a tournament of order n. By the induction hypothesis, $T - v$ contains a hamiltonian path, say $P: v_1, v_2, \ldots, v_n$. Now if (v, v_1) is an arc of T, then T contains the hamiltonian path $v, v_1, v_2, \ldots, v_n$. Also, if (v_n, v) is an arc of T, then T contains the hamiltonian path $v_1, v_2, \ldots, v_n, v$.

Suppose (v_1, v) is an arc of T. If all the vertices $v_1, v_2, \ldots, v_n$ are adjacent to v, then T contains a hamiltonian path because (v_n, v) is an arc of T. If not all the vertices $v_1, v_2, \ldots, v_n$ are adjacent to v, then there must be a vertex

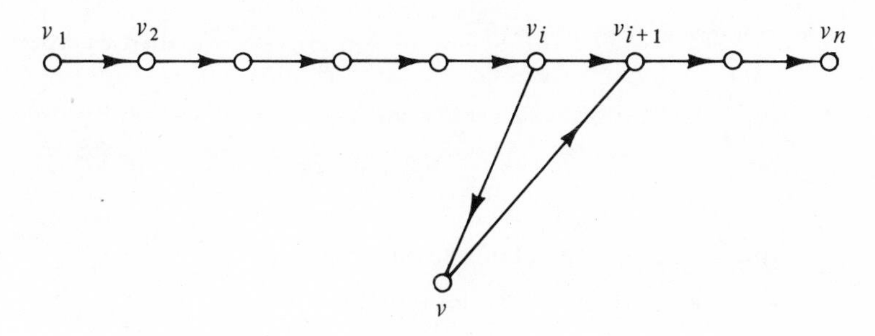

Figure 7.8

v_i, where $1 \leq i \leq n - 1$, such that (v_i, v) and (v, v_{i+1}) are arcs of T (see Figure 7.8). However, then

$$v_1, v_2, \ldots, v_i, v, v_{i+1}, v_{i+2}, \ldots, v_n$$

is a hamiltonian path. ∎

This result implies that at the conclusion of a round robin tournament, it is possible to rank the n teams as, say, $T_1, T_2, T_3, \ldots, T_n$, where team T_1 has defeated T_2, team T_2 has defeated T_3, etc.

Problem Set 7.2

10. How many non-isomorphic tournaments of order 4 are there? Draw them all.
11. Define: Tournament T_1 is isomorphic to tournament T_2.
12. Let U be a nonempty proper subset of the vertex set of a tournament T. Show that the removal of U from T results in a tournament.

13. If a graph G of size q has vertices $v_1, v_2, \ldots, v_p$, then we know that

$$\sum_{i=1}^{p} \deg v_i = 2q.$$

Suppose a digraph D of size q has vertices $v_1, v_2, \ldots, v_p$.

(a) What does $\sum_{i=1}^{p} \text{od } v_i$ equal?

(b) What does $\sum_{i=1}^{p} \text{id } v_i$ equal?

(c) What values do we get in (a) and (b) if D is a tournament?

14. (a) Give an example of a digraph D and two vertices u and v of D such that $d(u, v)$ is *not* defined.

(b) Under what conditions does a digraph D have the property that for every two vertices u and v of D, both $d(u, v)$ and $d(v, u)$ are defined?

15. Let u and v be distinct vertices of a tournament T such that both $d(u, v)$ and $d(v, u)$ are defined. Show that $d(u, v) \neq d(v, u)$.

16. (a) In Exercise 12 of Chapter 2 we defined a graph G (of order $p \geq 2$) to be *perfect* if no two of its vertices have equal degrees. What would be a reasonable definition of a "perfect digraph"? According to your definition, are there any perfect digraphs?

(b) What would be an appropriate definition of "regular digraph"? According to your definition, is there a regular tournament of order 4?

17. (a) If five teams play in a round robin tournament, show that it is possible for all five teams to tie for first place.

(b) If six teams play in a round robin tournament, show that it is *not* possible for all six teams to tie for first place.

*18. Theorem 7.2 states that if v is a vertex of maximum outdegree in a tournament T, then the distance from v to any other vertex of T is either one or two. Let n denote the number of vertices of T whose distance from v is one, and let m denote the number of vertices of T whose distance from v is two. Which of the following relations are possible: (i) $n > m$, (ii) $n = m$, (iii) $n < m$?

19. Theorem 7.3 implies that we may rank teams in a round robin tournament as $T_1, T_2, \ldots, T_n$, where team T_i defeats team T_{i+1} for $i = 1, 2, \ldots, n - 1$.
    (a) Give an example of a round robin tournament where such a ranking may give a true ordering of the teams.
    (b) Give an example of a round robin tournament where such a ranking may give a misleading ordering.

20. Prove that every tournament has at most one transmitter and at most one receiver.

7.3
Paired Comparisons and How to Fix Elections

There are many instances in our daily lives when we must make a choice among certain options. For example, a person may be planning to buy an automobile and must decide among a hardtop, sedan, van, or station wagon. Let us suppose that the preferences of a certain Mr. X are as follows (in order):

(1) van

(2) hardtop

(3) station wagon

(4) sedan

This list of preferences may also be given by the tournament T of Figure 7.9.

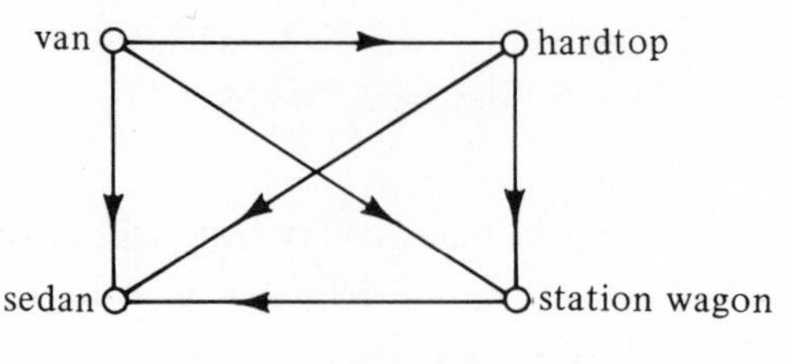

Figure 7.9

If we use a tournament T as a model of preferences, then the vertices represent the options and an arc is directed from vertex u to vertex v if the option corresponding to u is preferable to the option corresponding to v. If the tournament contains no cycles, then the choices made are consistent. For example, in the tournament T of Figure 7.9, we see that Mr. X prefers the van over the hardtop and prefers the hardtop over the station wagon. This suggests that Mr. X should prefer the van over the station wagon, which, indeed, he does.

As long as such tournaments of paired comparisons are consistent, we can reach some fair decisions. However, if a given tournament is inconsistent, then someone may be able to "swing" an interpretation in his favor by clever manipulation. We illustrate this.

Example 7.2

The Mathematics Department of a University has a decision-making committee with a place for one student to serve as a member. Three students, Frank (a freshman), John (a junior), and Gwen (a graduate student), are selected as a committee of three to determine which student should serve on this committee. They decide that the freshman, sophomore, junior, and senior classes, and the graduate students, should each nominate one student for membership on this faculty-student committee.

John volunteers to act as chairman of the three students, and begins discussions to decide which of the five students should be chosen. After some time, John observes that he, Frank, and Gwen have the following preferences:

John	*Frank*	*Gwen*
1. junior	1. freshman	1. graduate
2. senior	2. sophomore	2. senior
3. graduate	3. junior	3. sophomore
4. freshman	4. senior	4. freshman
5. sophomore	5. graduate	5. junior

These preferences can be shown by the tournaments in Figure 7.10 (page 164).

We can establish a cumulative tournament based on the tournaments (i.e., the preferences) of Frank, John, and Gwen. For example, both John and Frank prefer the freshman to the sophomore, while Gwen prefers the sophomore to the freshman. Thus, we conclude that the committee of Frank, John, and Gwen prefer the freshman to the sophomore. The cumulative preference tournament is given in Figure 7.11 (page 165).

John observes that the cumulative preference tournament is inconsistent. Knowing this, he plans his strategy. He proposes that since no two of the committee of three agree on the student representative, they should proceed by a process of elimination. They take a vote between the freshman and the sophomore, and the freshman wins. Next, they vote between the junior and the senior, and the junior wins. Hence, the sophomore and the senior have been eliminated. They next vote between the graduate student and the freshman, and the graduate student wins. Finally, they vote between the junior and the graduate student, and the junior (who just happens to be John's first choice) wins and is therefore chosen as student representative.

John:

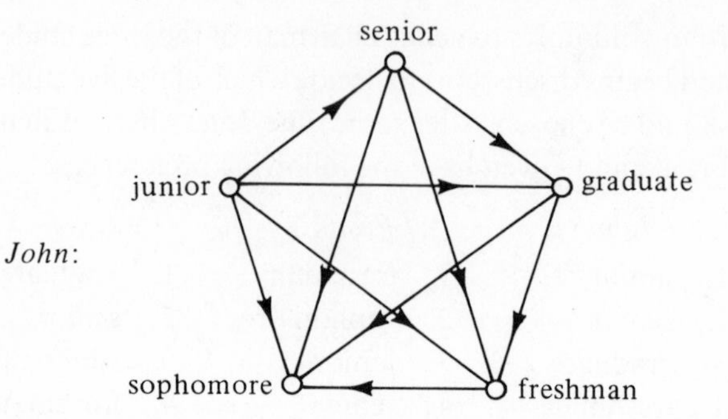

Frank:

senior
junior
graduate
sophomore
freshman

Gwen:

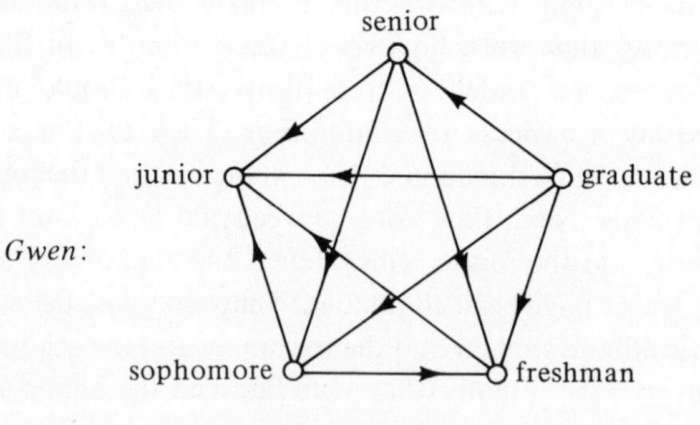

Figure 7.10

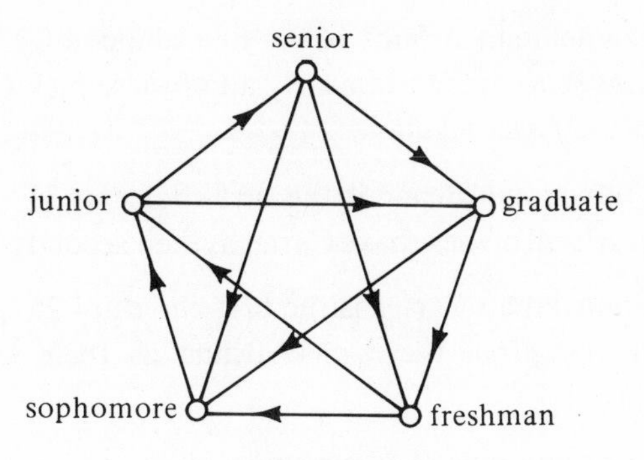

Figure 7.11

Example 7.3

A rarity for the town of Torrence! For the first time ever, three candidates are running for the office of mayor, namely Councilmen Adams, Beane, and Carter. The town elections officer does not know quite how to proceed with the election. Past records indicate that only one or two people have run for mayor in previous years, so there is no precedent to go by. The elections officer decides to ask for advice from the intelligent (and often unscrupulous) Professor Dixon, who teaches at the University only a few miles away. Professor Dixon says that before giving an opinion on this matter he would like to think about it for a few days. The Professor believes that the election will be close, so he decides to have two of his students conduct a poll of the voters of Torrence.

After two days of extensive polling, Professor Dixon's students report the following results:

(1) Of the 2500 registered voters of Torrence, two-thirds indicate that they plan to vote on Election Day.

(2) Councilman Adams is the first choice of 35% of those surveyed, and nearly all of these pick Carter as their second choice.

(3) Councilman Beane is the first choice of 30%, and nearly all of these have Carter as their second choice.

(4) Councilman Carter is the first choice of 25%, and nearly all of these pick Beane as their second choice.

(5) The remaining 10% are undecided.

Professor Dixon studies the results of this survey—and smiles. He sees how he can make some extra money. He calls Councilmen Adams, Beane, and Carter in turn and tells each one that for the right amount of money, he will see to it that *he* will be elected mayor. While waiting for their responses, Professor Dixon prepares the following three memos, of which only one will be given to the elections officer.

Memo A. The natural thing to do is to hold the election on Election Day, and declare the candidate with the highest number of votes the winner. Any kind of run-off election would cost the town additional money and would serve no purpose.

Memo B. The natural thing to do is to hold a preliminary election. The candidate with the lowest vote total is eliminated, and the final election is between the two top vote-getters from the preliminary election. This procedure assures that the newly elected mayor receives votes from a majority of the voters.

Memo C. The natural thing to do is to hold a preliminary election asking the voters to determine whom they least want as mayor. This person is then eliminated and the final election is held between the remaining two candidates.

Shortly after Professor Dixon prepares these memos, the police arrive at his home and arrest him for attempting to solicit bribes from the three councilmen. The memos are never delivered. The elections officer decides to hold a single election. As expected, Adams is the winner. The final vote is:

Adams	600
Beane	500
Carter	400

Problem Set 7.3

21. Give an example of a choice of options you have had recently which can be represented by a tournament.

22. With reference to Example 7.2, show how you could have altered the order of voting so that the student representative is
    (a) the freshman;
    (b) the sophomore;
    (c) the senior;
    (d) the graduate student.

23. Again, referring to Example 7.2, how *should* the student representative have been selected so as to reflect the preferences of Frank, John, and Gwen?

*24. A directed graph D is called *transitive* if whenever (u, v) and (v, w) are arcs of D, then (u, w) is also an arc of D. Show that a tournament T is transitive if and only if T has no cycles. (Thus, a tournament representing paired comparisons is consistent if and only if it is transitive.)

25. With reference to Example 7.3, who would have been elected mayor if Memo A had been delivered? Describe the election by means of a tournament of order 3.

26. In Example 7.3, who would have been elected mayor if Memo B had been delivered? Describe the election by means of a tournament of order 2.

27. In Example 7.3, who would have been elected mayor if Memo C had been delivered? Describe the election by means of a tournament.

28. With reference to Example 7.3:

    (a) In your opinion, in what order should the memos be listed to go from fairest to least fair?

    (b) In your experience, which memo would a community most likely follow?

29. Referring again to Example 7.3, construct a tournament T with vertices A, B, and C (Adams, Beane, and Carter) such that arc (u, v) belongs to T if, in a choice between candidates u and v, more voters prefer u than v. By this criterion, who should be elected mayor of Torrence?

30. In Example 7.3, who, *in your opinion*, should be elected mayor? Explain.

31. Ten sports broadcasters vote to decide the final rankings of college football teams by assigning points to the teams (10 to the best, 1 to the tenth). Nine of the broadcasters think State is best and Central is second. The tenth broadcaster (who went to Central) would like to see his alma mater ranked high. How could he affect the outcome?

Suggestions for Further Reading

The book by Harary, Norman, and Cartwright [4] gives a wealth of information on directed graphs and their applications, including discussions on voting. An interesting pair of articles on voting was written by Colman and Pountney [1,2]. Numerous facts about tournaments can be obtained from the survey article written by Harary and Moser [3]. A considerably more advanced book on tournaments has been written by Moon [5], who also wrote an interesting article [6] on voting.

[1] A. M. Colman and I. Pountney, "Voting paradoxes: a Socratic dialogue." *The Political Quarterly.* **46** (1975), pp. 186–190.

[2] A. M. Colman and I. Pountney, "Voting paradoxes: a Socratic dialogue (II)." *The Political Quarterly.* **46** (1975), pp. 304–309.

[3] F. Harary and L. Moser, "The theory of round robin tournaments." *American Mathematical Monthly.* **73** (1966), pp. 231–246.

[4] F. Harary, R. Z. Norman and D. Cartwright, *Structural Models: An Introduction to the Theory of Directed Graphs.* John Wiley & Sons, New York (1965).

[5] J. W. Moon, *Topics on Tournaments.* Holt. Rinehart, and Winston, New York (1968).

[6] J. W. Moon, "A problem on rankings by committees." *Econometrica*, **44** (1976), pp. 241–246.

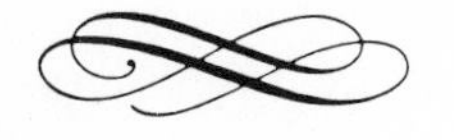

Chapter 8

Graphs and Social Psychology

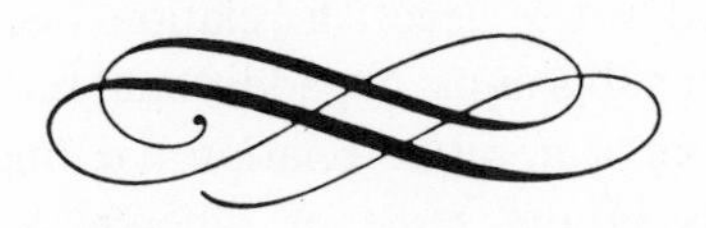

The investigation of relationships between people within a group is a branch of social psychology called group dynamics. Signed graphs and signed digraphs can serve as appropriate mathematical models in this area, with vertices representing the individuals of a particular group and positive and negative edges (or arcs) representing "positive" and "negative" relations. For the most part, graphs and digraphs have been used in social psychology primarily to lend clarity to situations rather than to aid in solving problems. In this chapter we consider some of the situations and problems that occur in social psychology.

8.1 The Problem of Balance

Suppose we have a group of people such that every two individuals are either (i) friendly toward each other; (ii) unfriendly toward each other; or (iii) indifferent toward each other. We refer to such a group of people with such relations between them as a *social system.* We may also use this term if "friendly" and "unfriendly" are replaced by the more general terms "positive relation" and "negative relation," respectively.

A social system is called *balanced* if all relations between people are positive, or if we can divide the group into two subgroups so that every positive relation occurs between individuals in the same subgroup, and every negative relation occurs between individuals in different subgroups. No additional restriction applies if neither a positive nor a negative relation (i.e., indifference) exists between individuals in the same or different subgroups. Some social psychologists have hypothesized a "tendency toward balance" in any social system, implying that an unbalanced system contains excessive stress or tension. The system tends to adjust so as to relieve this tension, by such acts as, for example, certain individuals within the group changing their points of view. Thus there is a tendency for the group to split into two factions such that within each faction the only relations are positive, and between factions the only relations are negative.

The concept of balance may be likened to a two-party system of government, where those people favoring a given party constitute the members of a faction. According to some social psychologists, then, such a system yields the greatest harmony within a social system.

We can represent a social system by a signed graph S whose vertices correspond to the individuals. A positive edge joins two vertices if there is a positive relation between the corresponding individuals, and a negative edge joins two vertices if there is a negative relation between the corresponding individuals. Indifference between two individuals is indicated by the lack of any edge joining the corresponding vertices. The following definition is now natural. A signed graph S is *balanced* if its vertex set can be partitioned into two subsets (one of which may be empty) so that each edge joining two vertices in the same subset is positive, while each edge joining vertices in different subsets is negative. It follows that balanced signed graphs are precisely those signed graphs which represent balanced social systems. In Figure 8.1, the signed graph S_1 is balanced, while the signed graph S_2 is unbalanced.

To verify that the signed graph S_1 of Figure 8.1 is balanced, consider the partition $V(S_1) = V_1 \cup V_2$, where $V_1 = \{u_1, u_2, u_3\}$ and $V_2 = \{u_4, u_5\}$. Every edge joining two vertices of V_1 or two vertices of V_2 is positive, while every edge joining a vertex of V_1 with a vertex of V_2 is negative. Therefore, S_1 is balanced.

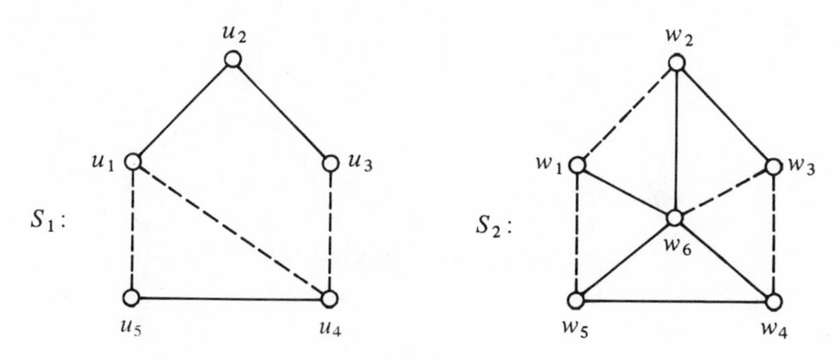

Figure 8.1

We show by using a contradiction argument that S_2 is unbalanced. Suppose S_2 *is* balanced. Then there exists a partition $V(S_2) = V_1 \cup V_2$ such that every edge connecting two vertices of V_1 or two vertices of V_2 is positive, while every edge between a vertex of V_1 and a vertex of V_2 is negative. Consider the vertex w_1. Without loss of generality, we may assume that $w_1 \in V_1$. Since w_1w_2 is a negative edge, $w_2 \in V_2$, and since w_1w_6 is a positive edge, $w_6 \in V_1$. However, $w_2 \in V_2$, $w_6 \in V_1$, but w_2w_6 is a positive edge, a contradiction. Therefore, S_2 is unbalanced.

Consider a social system consisting of three individuals A, B, and C such that no indifference exists between any two people. There are four such social systems; these are illustrated by the signed graphs of Figure 8.2. Figure 8.2(a) shows total harmony in the system. In Figure 8.2(b), individual C finds himself in the untenable position of agreeing with A and disagreeing with B, while A and B

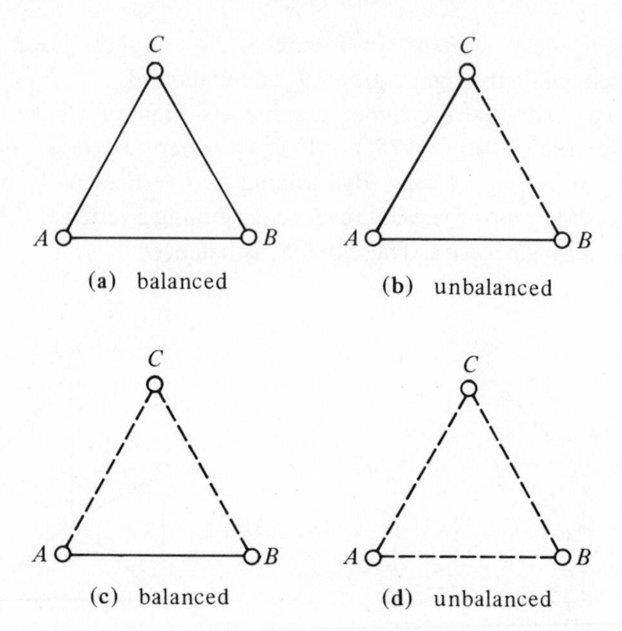

Figure 8.2

agree with each other. In Figure 8.2(c), *B* and *C* still disagree with each other, but now *C* disagrees with *A* as well as with *B*, while *A* and *B* still side with each other. There is no inconsistency with this arrangement. Figure 8.2(d) shows total disagreement among *A*, *B*, and *C*.

A path or cycle in a signed graph is called *positive* if it has an even number of negative edges, and is called *negative* otherwise. Thus, a path with no negative edges is positive.

We now prove two theorems giving necessary and sufficient conditions for a signed graph to be balanced.

Theorem 8.1

A signed graph S is balanced if and only if for every two vertices of S, all paths joining them have the same sign.

Proof

Let S be a balanced signed graph. We show that for each pair of vertices, all paths joining them have the same sign. If S contains only positive edges, then all paths are positive and the result follows immediately. If not all edges of S are positive, then we can partition the vertex set of S into two nonempty sets V_1 and V_2 such that an edge joining two vertices of V_i, $i = 1, 2$, is positive, while every edge joining a vertex of V_1 to a vertex of V_2 is negative. We now observe that any path joining two vertices of V_1 has an even number of negative edges, since whenever the path proceeds from V_1 to V_2 (using one negative edge), it must return to V_1 (thereby using another negative edge). By similar reasoning, every path joining two vertices of V_2 is positive, and every path joining a vertex of V_1 to a vertex of V_2 is negative.

Conversely, assume that S is a signed graph with the property that for each pair of vertices, all paths joining them have the same sign. We show that S is balanced. We can assume that S is connected, for if S is not connected and we can show that every component of S is balanced, it then follows that S itself is balanced. Let v be a vertex of S, and define the set V_1 to consist of v and all vertices x such that a positive path exists between x and v. If S has only positive edges (so that S is balanced), then V_1 contains all vertices of S.

Suppose that S has some negative edges, and let $V_2 = V(S) - V_1$. There can be no positive edge of the type v_1v_2, where $v_1 \in V_1$ and $v_2 \in V_2$, for this implies the existence of a positive path between v and v_2 or a negative path between v and v_1, contradicting the facts that $v_2 \notin V_1$ and $v_1 \in V_1$.

There can be no negative edge of the type uw, where $u, w \in V_i$, $i = 1, 2$, for suppose uw *is* negative. All paths from v to u have the same sign as those paths from v to w. If P is a path between v and u not containing w, then P followed by uw followed by w is a path between v and w having the opposite sign of P. If every path between v and u contains w, then there exists a v-w path P' not containing u. We can then produce a path P'' between v and u by taking P' followed by wu and then u. The paths P' and P'' have opposite signs. In either case, a contradiction arises, so uw must be positive.

Thus, $V_1 \cup V_2$ is an appropriate partition so that S is balanced. ▮

The following characterization of balanced signed graphs is now rather immediate.

Theorem 8.2

A signed graph S is balanced if and only if every cycle of S is positive.

Proof

Let S be a balanced signed graph, and suppose S has a negative cycle C. The cycle C therefore contains an odd number of negative edges. Let u and v be any two distinct vertices of C. The cycle C produces two edge-disjoint u-v paths, one necessarily containing an even number of negative edges, and the other containing an odd number of negative edges. This implies that there is a negative path as well as a positive path joining u and v, but this contradicts Theorem 8.1. Hence, every cycle of S is positive.

Assume now that S is a signed graph in which every cycle is positive. If S is not balanced, then by Theorem 8.1, there exist two vertices u and v and two u-v paths P' and P'' such that one path is positive and the other path is negative. It then follows that P' and P'' together produce a negative cycle, contradicting the hypothesis. Thus, S is balanced. ∎

Example 8.1

Consider a social system where friendliness or unfriendliness occurs between certain pairs of individuals. Assume there is a rumor which has two basic forms, one true and one false. Suppose anyone would tell the rumor to a friend in the same form he had received it, but would change the form if he were to pass the rumor to someone with whom he is unfriendly. If the system is balanced, each person will hear only one version of the rumor regardless of how it reached him; also, the person who started the rumor will hear it returned to him in the same form as he originally knew it.

Problem Set 8.1

1. Suppose we have a social system in which every two individuals dislike each other. Is this system balanced or unbalanced?
2. Suppose we have a social system in which every two individuals are indifferent toward each other. Is this system balanced or unbalanced?
3. Let S be a signed graph with no cycles whatsoever. Is it possible to determine whether S is balanced or unbalanced? Explain.

4. Show that in Figure 8.3 the signed graph S_1 is balanced and the signed graph S_2 is not balanced.

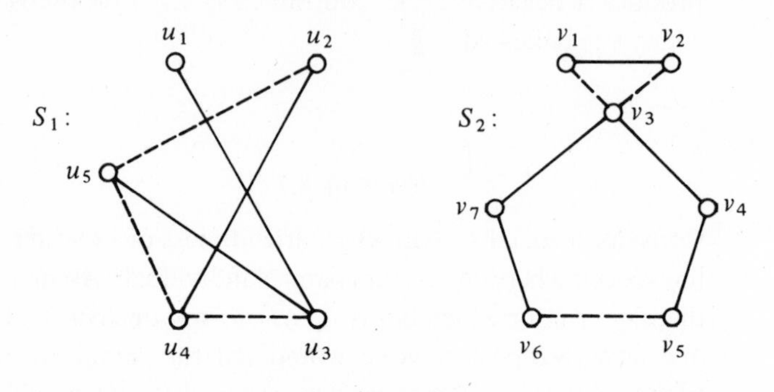

Figure 8.3

*5. Prove that if $S_1, S_2, \ldots, S_n$ are connected balanced signed graphs, then the disconnected signed graph S, having $S_1, S_2, \ldots, S_n$ as components, is balanced.

*6. Let S_1 and S_2 be two connected balanced signed graphs, and let S be the connected signed graph obtained by "identifying" a vertex of S_1 with a vertex of S_2 (i.e., a vertex of S_1 is combined with a vertex of S_2 to form one vertex of S). Prove that S is balanced.

*7. In the proof of Theorem 8.2, supply the details of the fact that the paths P' and P'' produce a negative cycle.

8. What signed graph would represent a "love triangle"? Is this signed graph balanced or unbalanced? If you answered "unbalanced," how would you interpret a tendency toward balance? If you answered "balanced," what assumptions are you making?

9. A city suburb contains a very closely knit neighborhood. A new family of lower social status moves into this neighborhood. At first, all of the original neighbors have nothing to do with

this family, but then one of the original families makes friends with the newcomers. What might happen in the rest of this neighborhood?

8.2
The Problem of Clustering

Figure 8.2(d) illustrated a social system containing three individuals in which only negative relations exist. By definition, such a system is unbalanced, implying that it is unnatural and susceptible to inner tension. However, we must emphasize that this is only a point of view among *some* social psychologists. A related but alternate theory is to anticipate a "clustering" of people into factions or *clusters* (not necessarily two) where the only relations occurring between individuals in the same cluster are positive, and the only relations occurring between individuals in different clusters are negative. There may be indifference between some individuals. Social systems possessing this property are called *clusterable.* Clusterable social systems are therefore generalizations of balanced social systems. The signed graph of Figure 8.2(d) represents a clusterable social system with three clusters.

It is now convenient to define clusterable signed graphs. A signed graph S is said to be *clusterable* if its vertex set can be partitioned into subsets, called *clusters,* so that every positive edge joins vertices within the same subset and every negative edge joins vertices in different subsets.

We can now present a characterization of clusterable signed graphs.

Theorem 8.3

A signed graph S is clusterable if and only if S contains no cycle with exactly one negative edge.

Proof

Assume that S is a clusterable signed graph. We show that no cycle of S has exactly one negative edge. Let C be a cycle of S. If C contains vertices from a single cluster, then all edges of C are positive. If C contains vertices from two or more clusters of S, then C contains at least two edges joining different clusters, i.e., at least two negative edges. However, in either case, C does not contain exactly one negative edge.

Conversely, let S be a signed graph containing no cycle with exactly one negative edge. Let u and v be any elements of the vertex set V of S. Now we define a relation R on V such that $(u, v) \in R$ if $u = v$ or if u and v are joined by a path whose edges are all positive. The relation R is reflexive, symmetric, and transitive; that is, R is an equivalence relation. Recall that the equivalence class $[x]$, where $x \in V$, consists of all vertices related to x. By Theorem A.2 in the Appendix, the equivalence classes determined by R form a partition of V. We show that these equivalence classes are clusters of S, so that S is clusterable.

There can be no positive edge joining vertices u and v in different equivalence classes, for if there were, we would have a u-v path all of whose edges (actually only one edge) are positive, so that u is related to v, and u and v belong to the same equivalence class. Also, there can be no negative edge joining vertices y and z in the same equivalence class, because since y and z belong to the same equivalence class, there exists a y-z path P all of whose edges are positive. However, the path P together with the negative edge yz produces a cycle having exactly one negative edge, contradicting our definition of S. Therefore, the equivalence classes serve as clusters and S is a clusterable signed graph. ▮

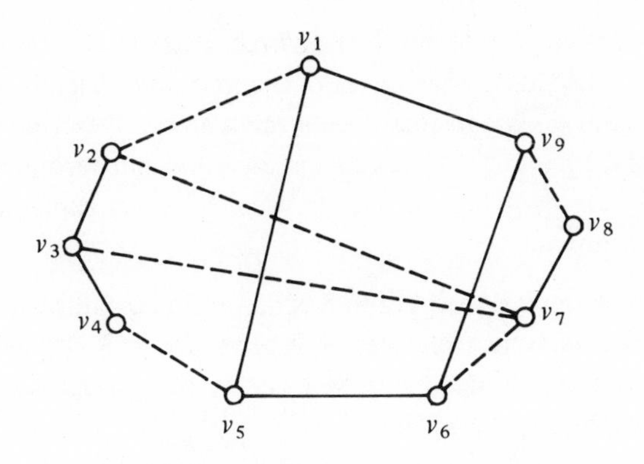

Figure 8.4

Figure 8.4 shows a signed graph which is clusterable but not balanced. The clusters of this signed graph are

$$V_1 = \{v_1, v_5, v_6, v_9\}, \quad V_2 = \{v_2, v_3, v_4\},$$

and

$$V_3 = \{v_7, v_8\}.$$

Problem Set 8.2

10. Show that the signed graph of Figure 8.4 is clusterable by verifying that the given clusters possess the desired properties.

*11. Show that if S is a disconnected signed graph and every component of S is clusterable, then S is clusterable. Determine the minimum number of clusters of S in terms of the minimum number of clusters in the components of S.

12. Give a real-life example of a social system in which clustering is more appropriate than balance.

13. A married couple obtain a divorce. Prior to the divorce, many of the friends and relatives of the man were friends with many of the friends and relatives of the woman. Describe a possible social system (in terms of balance and clustering) consisting of these people after the divorce. Give an explanation with your answer.

14. A *clique* in a signed graph S is defined as a complete subgraph S_0 of S such that all edges of S_0 are positive. A clique in a social system may be described as a closely knit group of people with common interests.
    (a) What is the connection between cliques and clusters in a clusterable signed graph?
    (b) Are you aware of cliques among people you know? Are you a member of a clique?

8.3
Graphs and Transactional Analysis

We have already noted that graphs are quite useful in visualizing certain situations occurring in social psychology; however, theorems about graphs have not been particularly important in this area. Another pictorial use of graphs in social psychology is explored in this section.

In social psychology, a *transaction* consists of two parts: the transactional stimulus and the transactional response. In any gathering of two or more people, eventually one of them will recognize the presence of the others. (This could take the form of the person speaking, smiling, nodding, etc.) This is the transactional stimulus. A reaction by one of the other people to this stimulus is called the transactional response. As the term suggests, transactional analysis is the analysis or examination of transactions, particularly between a specific individual and other people.

The psychologists and psychiatrists who deal in transactional analysis believe that each person exists in one of three states, referred to as the Parent, the Adult, and the Child. In terms of graphs, it is customary in these circumstances to represent an individual as a "vertex triple" (as shown in Figure 8.5), with P, A, and C denoting the Parent, the Adult, and the Child, respectively.

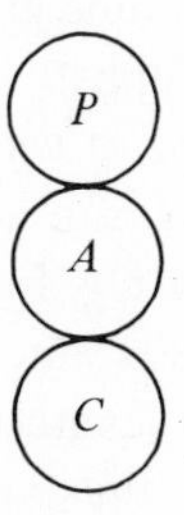

Figure 8.5

The Parent consists of the recordings or recollections of the external stimuli the individual experienced in his early life, ordinarily that period prior to attending school. Since the most pronounced experiences of this time are his parents (or guardians), the term Parent is employed. The Parent state includes the numerous rules laid down by the parents, consisting of "what to do" and "what not to do." The significant feature of these recordings is that they are retained as the truth.

The Child consists of recordings of the responses to experiences an individual encountered in his early life. Most of these reactions take the form of feelings, since oral communication is limited in the early years. There are many demands put on a young child, and the child does not understand why these requirements are placed on him. The Child often contains feelings of anger, rejection, and frustration. However, it often contains such positive reactions as curiosity.

At some point in life, the individual has more control of his responses and is able to make decisions on his own, rather than do things the way it was shown to him or the way he wished it could be. The Adult, therefore, is the state that makes decisions based on various information gathered. This differs from the Parent, where reactions are by imitation, or the Child, where reaction is rapid and unthinking.

We are primarily concerned with relations between individuals and the states they may be in at a given instant. We shall illustrate this by examples, at the same time giving the appropriate model in terms of graphs (actually digraphs).

The most enjoyable type of transaction is that referred to as complementary. In this kind of transaction, two individuals A and B are in states α and β, respectively. Furthermore, A addresses the β state of B, and B addresses the α state of A. We represent this transaction as a (directed) cycle of length two between two states in the graphical model. We consider this type first.

Example 8.2

Let us examine three conversations going on between certain pairs of individuals at an airport. First, two men begin a conversation.

Mr. A: Excuse me. What time do you have?

Mr. B: It's 10:40. Are you taking this flight?

Mr. A: Yes, I'm going to San Francisco.

Mr. B: I'm going to Denver. I believe it's the only stop between here and San Francisco.

Mr. A: That's right. I take this flight quite regularly.

The transaction between Mr. A and Mr. B is an Adult–Adult transaction (see Figure 8.6).

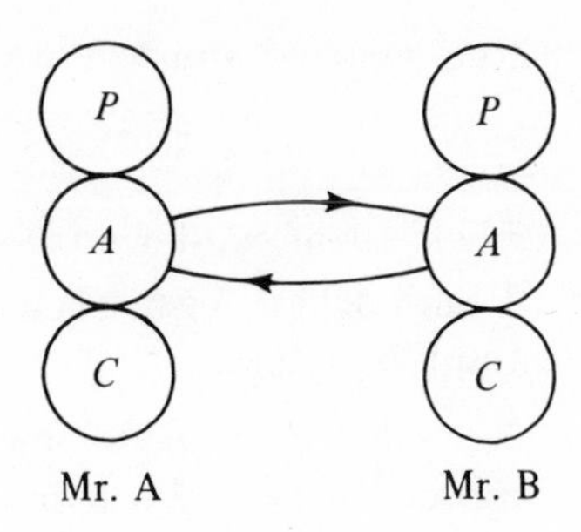

Figure 8.6

A gentleman and a lady begin a conversation.

Ms. C: Why don't they let us on the plane? I want a seat next to the window.

Mr. D: We'll probably be late. I've never taken a plane that has left on time.

Ms. C: I haven't either. I don't know why they bother to list departure times.

Mr. D: If I ran my business this way, I'd be broke.

This is an example of a Parent–Parent transaction and is illustrated in Figure 8.7.

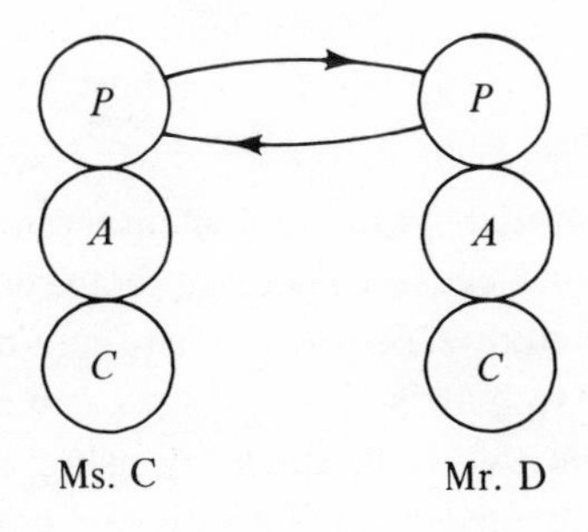

Figure 8.7

Next we listen to a husband and wife waiting for their flight.

Mrs. E:	Don't shake so much. It's embarrassing.
Mr. E:	I can't help it. I get this way whenever I think of flying.
Mrs. E:	It's safer in a plane than in a car—especially the way you drive.
Mr. E:	Don't yell at me! You make me nervous.
Mrs. E:	Stop whimpering!

This is an example of a Parent–Child transaction (see Figure 8.8).

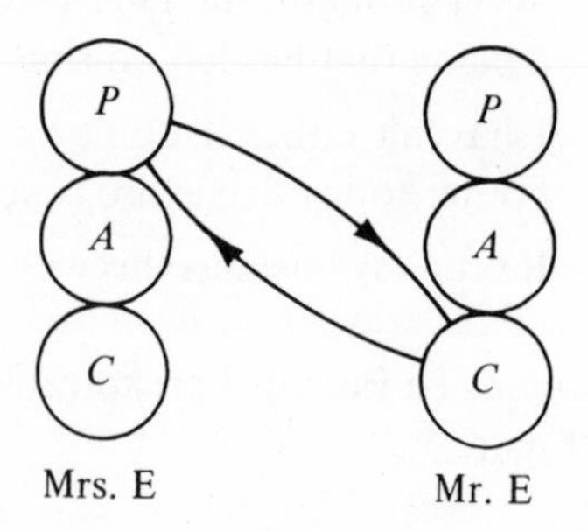

Figure 8.8

We next consider noncomplementary transactions. These types of transactions cause difficulties between people, result in communication breakdowns, and can lead to a need for therapy by a psychologist or psychiatrist. In representing a noncomplementary transaction between two individuals by using a digraph, no cycle is created: only a (directed) path of length two, or two paths of length one each.

Example 8.3

We consider three situations which result in non-complementary transactions. In each case, two dormitory roommates are having a discussion that quickly leads to a lack of communication.

In the first case, both students are in their Parent state and address the other individual's Child state (see Figure 8.9).

Mr. A: Why don't you pick up your clothes? This room's a mess.

Mr. B: I pay half the rent—I can do what I want. Your friends are what make this room a mess.

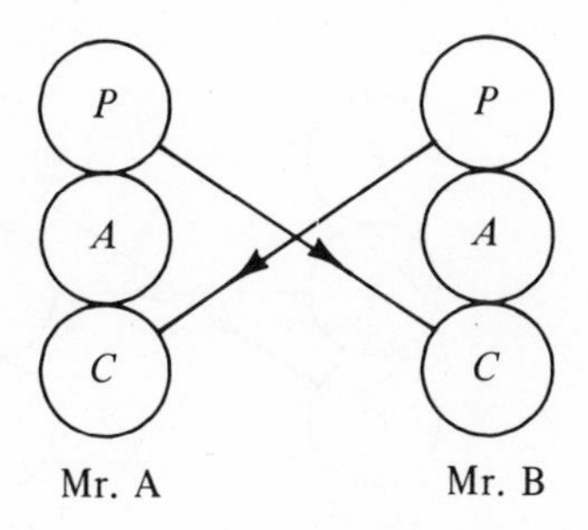

Figure 8.9

In the next room, one student is in the Adult state and the other is in the Child state (see Figure 8.10).

Mr. C: What are you studying tonight?

Mr. D: Math! Stupid math! Why don't they ever write a math book that makes sense? Who needs math, anyway?

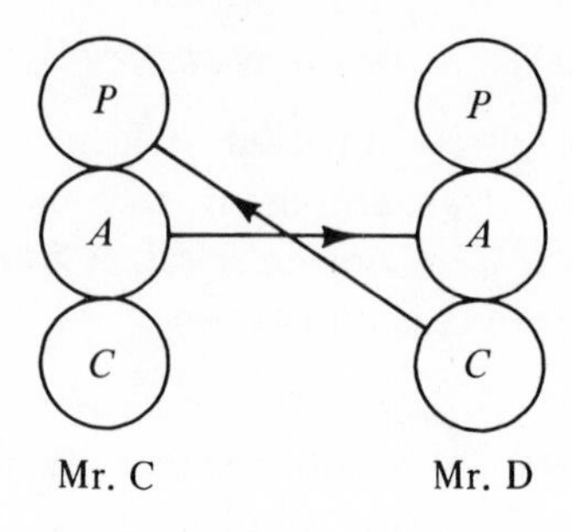

Figure 8.10

In the final situation, one student is in his Adult state while the other is in his Parent state (see Figure 8.11).

Mr. E: I've decided to major in math.

Mr. F: What makes you think you're smart enough to major in anything?

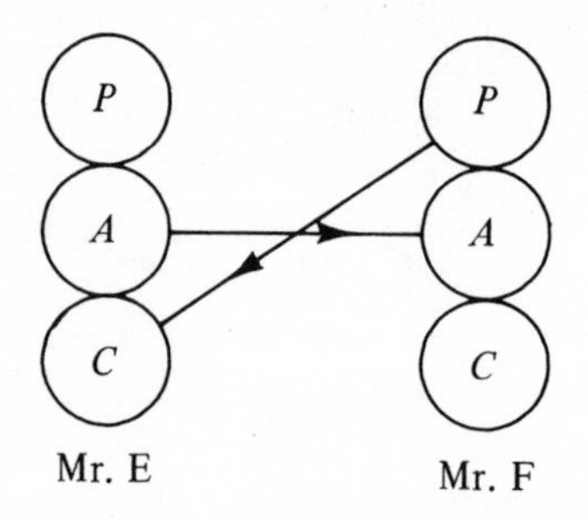

Figure 8.11

Problem Set 8.3

15. Give two responses for each of the following statements, one response causing a complementary transaction and the other

a noncomplementary transaction. Draw the appropriate digraphs.

(a) *Mr. A*: This dormitory food is terrible. I'm not eating it, no matter what.

(b) *Mr. C*: Girls aren't any good in mathematics.

(c) *Miss E*: I dropped my pen behind your desk. Would you pick it up for me, please?

16. Give examples of two transactions such that one can be represented by the digraph in Figure 8.12(a) and the other can be represented by Figure 8.12(b).

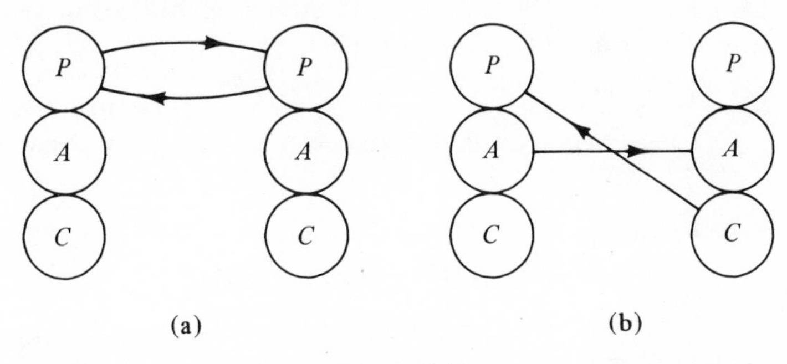

Figure 8.12

17. A hockey player telephones a girl, asking her for a date. Give four sample conversations: two complementary and two noncomplementary transactions. Draw the corresponding digraphs.

Suggestions for Further Reading

Heider [5] is credited with developing much of balance theory, while Davis [3] initiated the idea of clustering. Two other papers of interest on these topics are by Cartwright and Harary [2] and by Riley [6].

Berne is considered the originator of transactional analysis (see [1], for example); however, a popular book on this subject has been written by Harris [4].

[1] E. Berne, *Games People Play*. Grove Press, New York (1964).

[2] D. Cartwright and F. Harary, "Structural balance: a generalization of Heider's theory." *Psychological Review*, **63** (1956), pp. 277–293.

[3] J. A. Davis, "Clustering and structural balance in graphs." *Human Relations*, **20** (1967), pp. 181–187.

[4] T. A. Harris, *I'm OK—You're OK: A Practical Guide to Transactional Analysis*. Harper and Row, New York (1967).

[5] F. Heider, "Attitudes and cognitive organization." *Journal of Psychology*, **21** (1946), pp. 107–112.

[6] J. E. Riley, "An application of graph theory to social psychology." *The Many Facets of Graph Theory* (G. Chartrand and S. F. Kapoor, eds.). Springer-Verlag, Berlin (1969), pp. 275–280.

Chapter 9

Planar Graphs and Coloring Problems

Two major concepts in graph theory are planarity and chromatic number. In this chapter we consider two problems whose solutions illustrate these concepts. We then discuss a problem which combines both ideas: the famous Four Color Problem.

9.1 The Three Houses and Three Utilities Problem: An Introduction to Planar Graphs

We begin by stating the problem which concerns us in this section.

Chapter 9
Planar Graphs and Coloring Problems

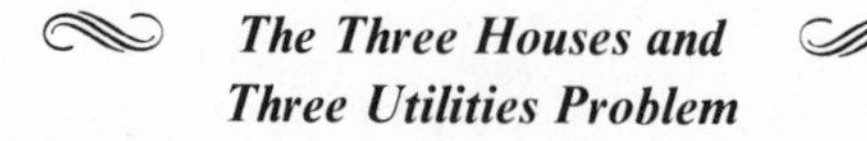

The Three Houses and Three Utilities Problem

SUPPOSE WE HAVE three houses and three utility outlets (electricity, gas, and water) situated as shown in Figure 9.1. Is it possible to connect each utility with each of the three houses without the lines or mains crossing?

Figure 9.1 shows one way we might try to connect each utility with each house; however, this procedure evidently fails, since there appears to be no way of connecting the third house to the water outlet without crossing some line or main. This implies that either this procedure for connecting the houses and utilities is incorrect, or there is no way of doing it.

To find a solution to this problem, we once again turn to graph theory. We can represent the situation described in the Three Houses and Three Utilities Problem by a graph whose vertices correspond to the houses and utilities, and where an edge joins two vertices if and only if one vertex denotes a house and the other vertex denotes a utility. This bipartite graph is denoted by $K(3, 3)$,

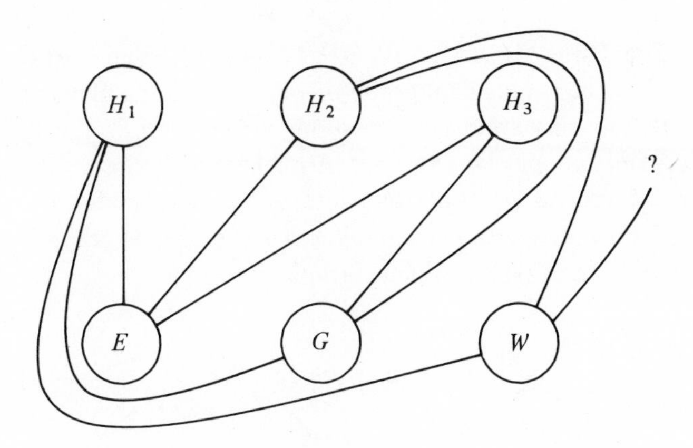

Figure 9.1

indicating that the vertex set is partitioned into two sets of three vertices each, and all allowable edges are drawn. The graph $K(3, 3)$ is shown in Figure 9.2.

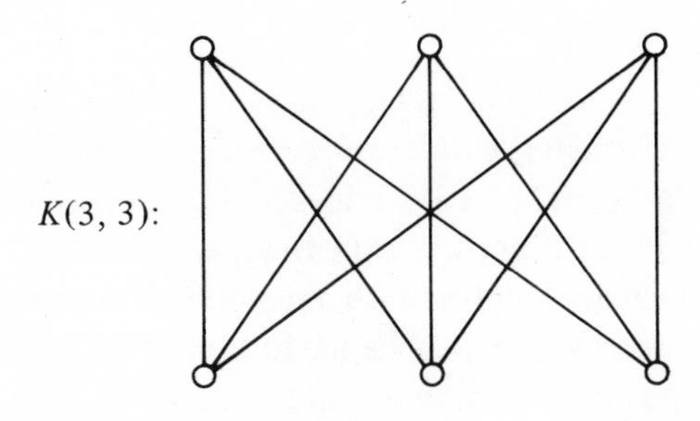

Figure 9.2

The concept we are primarily interested in is planar graphs. A *planar graph* is a graph that can be drawn in the plane in such a way that no two edges intersect except at a vertex. For example, the graph G of Figure 9.3(a) is drawn with intersecting edges, but it is a planar graph because G *can* be drawn, as in Figure 9.3(b), so that no edges intersect.

The Three Houses and Three Utilities Problem can now be restated in the language of graph theory: Is $K(3, 3)$ a planar graph?

Before answering this question, we need to establish a few basic results about planar graphs. A planar graph already drawn in the plane so that no two edges intersect is referred to as a *plane graph*. Thus, the graph of Figure 9.3(a) is not a plane graph, but the graph of Figure 9.3(b) is a plane graph.

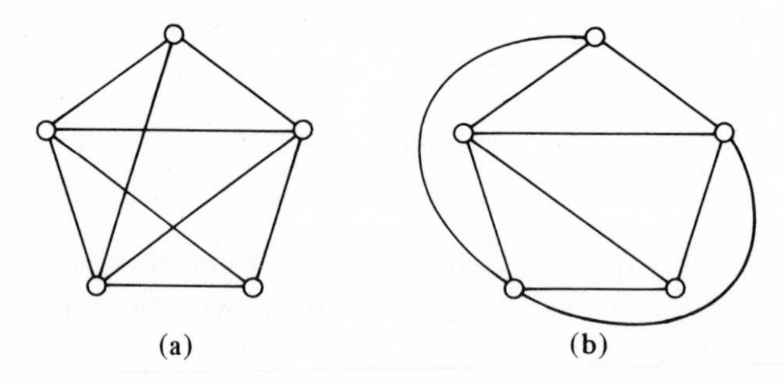

Figure 9.3

Let G be a plane graph and consider the parts of the plane remaining after we remove the edges and vertices of G. These connected pieces of the plane are called *regions* of G. The vertices and edges of G which are incident with a region R make up the *boundary* of R. Let us illustrate these concepts. In Figure 9.4 the graph G_1 has three regions, G_2 has one region, and G_3 has six regions. The boundary of the region R_1 of G_3 consists of the vertices v_2, v_3, and v_4 and the edges v_2v_3, v_2v_4, and v_3v_4; the boundary of R_6 consists of the vertices v_1, v_2, v_3, v_5, v_6, and v_7 and the edges $v_1v_2, v_2v_3, v_3v_5, v_5v_6$, v_6v_7, and v_2v_7. The region R_6 is called the *exterior region* of G_3. Every plane graph always has exactly one exterior region.

The plane graph G_1 of Figure 9.4 has $p = 4$ vertices, $q = 5$ edges, and $r = 3$ regions; G_2 has $p = 5$, $q = 4$, and $r = 1$; while G_3 has $p = 7, q = 11$, and $r = 6$. Observe that in all three cases, $p - q + r = 2$. This is no mere coincidence, as we now show.

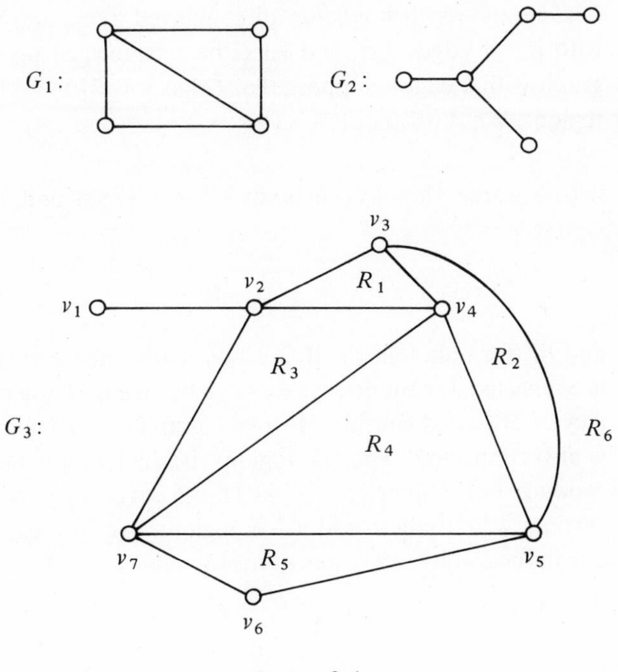

Figure 9.4

Theorem 9.1

Let G be a connected plane graph with p vertices, q edges, and r regions. Then

$$p - q + r = 2.$$

Proof

We use induction on q. If $q = 0$, then $p = 1$ and $r = 1$, so

$$p - q + r = 1 - 0 + 1 = 2,$$

and the result is true.

Assume the result holds for all connected plane graphs with $k - 1$ edges, say, and let G be a connected plane graph with k edges. Suppose that G has p vertices and r regions. We show that $p - k + r = 2$.

If G is a tree, then by Theorem 4.2, $p = k + 1$ and, of course, $r = 1$, so

$$p - q + r = (k + 1) - k + 1 = 2$$

and the formula follows. If G is not a tree, then since G is connected, G contains cycles. Let e be an edge lying on a cycle of G, and consider the plane graph $G - e$ (which is also connected). The two regions incident with e in G produce one region in $G - e$. Therefore $G - e$ has p vertices, $k - 1$ edges, and $r - 1$ regions. We can apply our induction hypothesis to conclude that

$$p - (k - 1) + (r - 1) = 2$$

or

$$p - k + r = 2,$$

which is what we wanted to prove. ∎

Theorem 9.1 is a classical theorem of graph theory; in fact, the result is due to Euler. We can now verify another useful result, whose proof involves a rather tricky counting argument.

Theorem 9.2

Let G be a connected planar graph with p vertices and q edges, where $p \geq 3$. Then

$$q \leq 3p - 6.$$

Proof

We first note that the result is true for $p = 3$, since every graph of order three has size at most three. Hence, we assume $p \geq 4$. We draw the graph G as a plane graph, and denote the number of regions of G by r. For each region R of G, we determine the number of edges lying on the boundary of R, and then sum these numbers over all regions of G. We call this number N. Since there are at least three edges belonging to the boundary of every region, it follows that $N \geq 3r$. On the other hand, the number N counts every edge of G once or twice; that is, $N \leq 2q$. Hence,

$$3r \leq N \leq 2q \quad \text{or} \quad 3r \leq 2q.$$

This says that $-r \geq -2q/3$. Now by Theorem 9.1, we know that

$$p - q + r = 2, \quad \text{or} \quad p = q - r + 2.$$

Therefore,

$$p = q - r + 2 \geq q - 2q/3 + 2 = q/3 + 2$$

so $p \geq q/3 + 2$. Thus, $q \leq 3p - 6$. ∎

Intuitively, Theorem 9.2 states that a planar graph of order p cannot have too many edges. We now use an argument similar to the proof of Theorem 9.2 to arrive at our main goal.

Theorem 9.3
Solution of the Three Houses and Three Utilities Problem

The graph $K(3, 3)$ *is not planar.*

Proof

We prove this result by contradiction. Suppose $K(3, 3)$ is planar. Then we can draw $K(3, 3)$ as a plane graph having r regions. We sum the number of edges lying on the boundary of each region over all r regions of $K(3, 3)$, and denote this number by N. Since no edges join vertices within each vertex subset of $K(3, 3)$, the graph contains no triangles, so it follows that $N \geq 4r$. On the other hand, N counts each edge at most twice, so $N \leq 2q = 18$. Therefore,

$$4r \leq 18 \quad \text{or} \quad r \leq 9/2.$$

However, by Theorem 9.1, $p - q + r = 2$, so

$$6 - 9 + r = 2 \quad \text{or} \quad r = 5.$$

This produces the desired contradiction. ▮

We can now see that it is impossible to connect each utility with each of the three houses without lines or mains crossing.

As an application of Theorem 9.2, we show that the complete graph K_5 is not planar.

Theorem 9.4

The graph K_5 is not planar.

Proof

The graph K_5 has $p = 5$ and

$$q = p(p - 1)/2 = 10.$$

Since $3p - 6 = 9$, we have $q > 3p - 6$. Therefore, by Theorem 9.2, K_5 is not planar. ▮

The nonplanar graphs K_5 and $K(3, 3)$ play a major role in the theory of planar graphs. We discuss this briefly.

By a *subdivision of a graph G*, we mean a graph obtained from G by inserting vertices (of degree two) into the edges of G. For the graph G of Figure 9.5, the graph H is a subdivision of G, while F is not a subdivision of G.

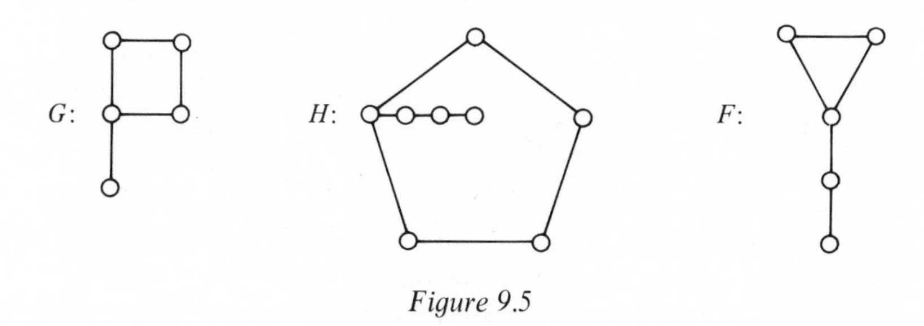

Figure 9.5

We can now state a well-known theorem from graph theory. However, the proof is too complicated to present here.

Theorem 9.5

A graph G is planar if and only if G contains no subgraph isomorphic to K_5 or $K(3, 3)$ or any subdivision of K_5 or $K(3, 3)$.

Although we introduced planar graphs by means of a problem which might be of interest, the problem is not considered to be of any real importance. However, planar graphs are intimately associated with printed electronic circuits, where vertices represent terminals and an edge joins two vertices if a conducting path connects the two corresponding terminals. We do not wish two paths to cross, for this would create a short circuit.

Problem Set 9.1

1. Draw the graphs $K(2, 3)$, $K(2, 4)$, $K(2, 5)$, and $K(4, 4)$.
2. Show that the graphs G_1 and G_2 of Figure 9.6 are planar.

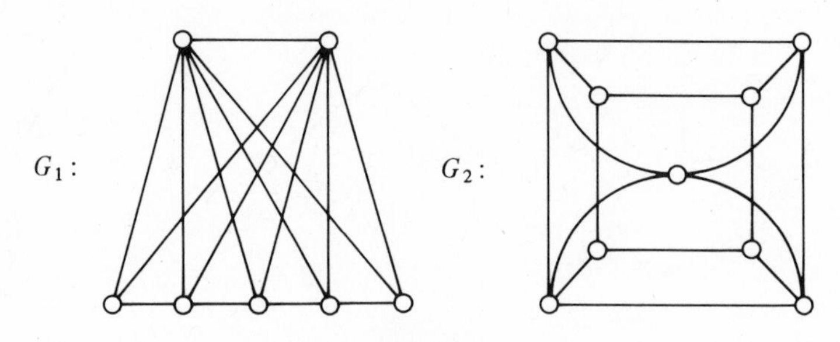

Figure 9.6

3. For the plane graph G of Figure 9.7, determine the vertices and edges of the boundary of region R and the boundary of the exterior region.

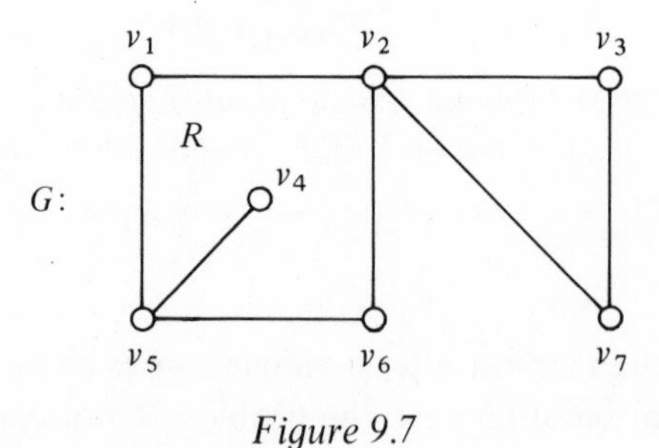

Figure 9.7

4. Determine plane graphs G_1' and G_2' isomorphic to G_1 and G_2, respectively, of Figure 9.6. Use Theorem 9.1 to determine the number of regions of G_1' and G_2'.

5. Let G be a connected planar graph. Show that no matter how G is drawn as a plane graph, it always has the same number of regions.

6. Figure 9.8 shows a tetrahedron and a cube. These are examples of polyhedra, which are geometric solids whose faces are all identical. Let V denote the number of vertices of a polyhedron, and let E and F denote the number of edges and number of faces, respectively. Determine $V - E + F$ for the tetrahedron and the cube. Compare this with Theorem 9.1.

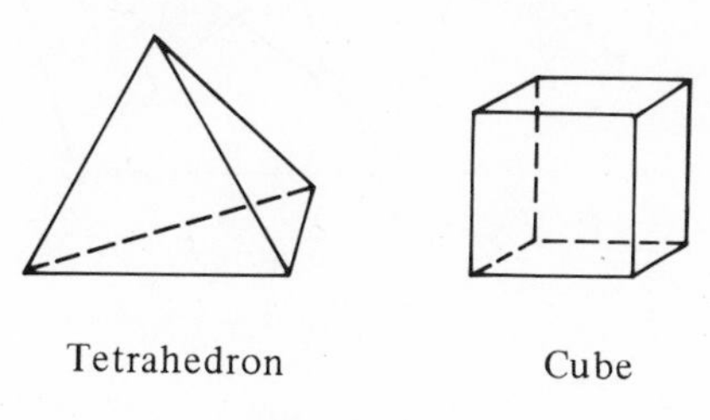

Tetrahedron Cube

Figure 9.8

7. (a) Give an example of a connected (p, q) planar graph such that equality holds in Theorem 9.2.

   (b) Give an example of a connected (p, q) planar graph such that the strict inequality holds in Theorem 9.2.

*8. Show that Theorem 9.2 is true even if G is disconnected.

*9. True or false? If G is a connected (p, q) graph and $q = 3p - 6$, then G is planar. Explain.

10. Suppose the Three Houses and Three Utilities Problem were the Five Houses and Two Utilities Problem. What would the solution be? What if it were the n Houses and Two Utilities Problem? What if it were the n Houses and Three Utilities Problem?

11. Which complete graphs are planar?

12. Show that $K(2, 2)$ is a subdivision of K_3.

*13. Figure 9.9 shows a famous graph called the Petersen graph. Is the Petersen graph planar? (*Hint*: Try to make use of Theorem 9.5.)

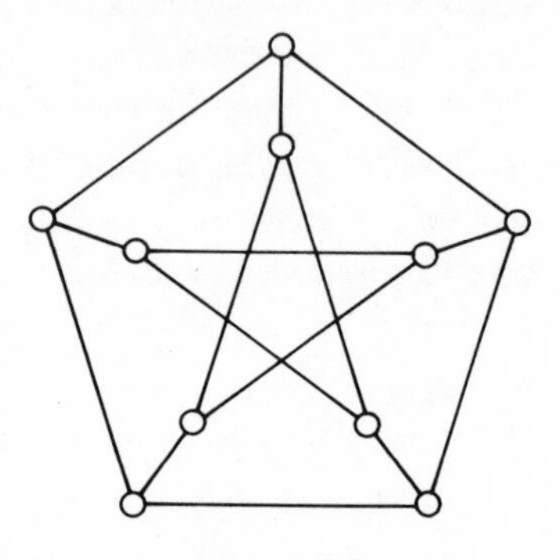

Figure 9.9

9.2
A Scheduling Problem: An Introduction to Chromatic Numbers

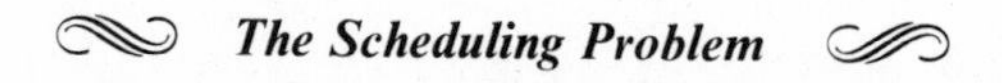

The Scheduling Problem

SUPPOSE YOU ARE A department chairman at a college, and one of your responsibilities is to set up the schedule of courses offered for the next semester. If the college is not too large, it might very well be practical to have the students indicate which courses they plan to take. As you set up the schedule, you must, of course, be careful not to have two courses meet at the same time if a student plans to take both courses. On the other hand, it is most convenient to set up a schedule requiring the fewest number of time periods during the day, since then it might be possible to eliminate classes meeting at unpopular times (such as the time this class meets). The question is: What is the minimum number of hours needed for such a schedule?

We first show how this problem is related to graphs. We construct a graph G_0 by associating a vertex of G_0 with each class in the schedule. We join two vertices of G_0 by an edge if and only if some student is a member of both corresponding classes. We now interrupt our consideration of the graph G_0 to introduce a new graph theoretic concept. We shall soon show a relationship between the two ideas.

By a *coloring* of a graph G, we mean the assignment of colors (which are simply the elements of some set) to the vertices of G, one color to each vertex, so that adjacent vertices are assigned different colors. An *n-coloring* of G is a coloring of G using n colors. Figure 9.10 shows a 5-coloring of the graph G_1 where the colors are denoted by numbers, as well as a 4-coloring of the graph G_2.

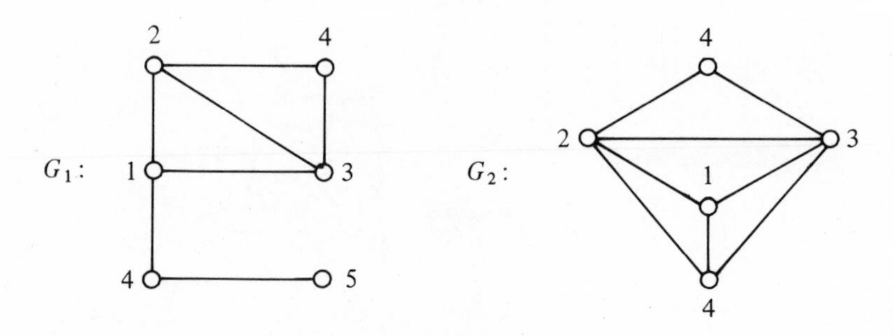

Figure 9.10

Observe that it is always possible to give a p-coloring to a graph G of order p. It is a more interesting (and more challenging) problem to determine those numbers n, where $n < p$, for which an n-coloring of G exists. Both a 4-coloring and a 3-coloring exist for the graph G_1 of Figure 9.10; however, there is no n-coloring, where $n < 4$, for the graph G_2 of Figure 9.10.

We now introduce the main concept of this section. The *chromatic number* of a graph G is the minimum value n for which an n-coloring of G exists. The chromatic number of G is denoted $\chi(G)$ ("χ" is the Greek letter chi). Thus, for the graphs of Figure 9.10 we have $\chi(G_1) = 3$ and $\chi(G_2) = 4$. Note that although we have stated the chromatic numbers of these two graphs, we have not given any verification of these facts.

We now prove that $\chi(H) = 4$ for the graph H of Figure 9.11. First, it is clear that $\chi(H) \leq 4$ since Figure 9.11(a) gives a 4-coloring of H. It will follow that $\chi(H) = 4$ if we can show there is no 3-coloring of H. Suppose there is a 3-coloring of H, and denote the colors by 1, 2, and 3. Suppose the vertices of H are labeled as in Figure 9.11(b).

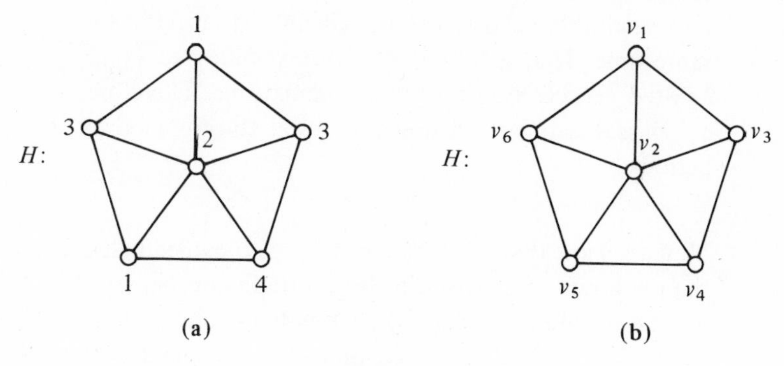

Figure 9.11

Now the vertices v_1, v_2, and v_3 form a triangle; therefore, three colors are required to color these vertices. Without loss of generality, we assign color 1 to v_1, color 2 to v_2, and color 3 to v_3. Since v_4 is adjacent to both v_2 and v_3, we must assign color 1 to v_4. (Recall that we are assuming only colors 1, 2, and 3 are necessary.) By the same reasoning, we must assign the color 3 to v_5. However, now v_6 is adjacent to a vertex colored 1, a vertex colored 2, and a vertex colored 3. This gives us a contradiction. Hence, there exists no 3-coloring of H.

We now return to our scheduling problem and the resulting graph G_0.

Theorem 9.6

The minimum number of hours required for the schedule of classes in our scheduling problem is $\chi(G_0)$.

Proof

Suppose $\chi(G_0) = m$, and suppose the colors used in coloring G_0 are $1, 2, \ldots, m$. We first claim that all classes can be scheduled in m one-hour time periods. In order to see this, consider all those vertices colored 1, say, and the classes corresponding to these vertices. Since no two vertices colored 1 are adjacent, no two classes corresponding to such vertices contain the same student. Hence, all these classes can be scheduled to meet at the same time. That is, all classes corresponding to same-colored vertices can meet at the same time. Therefore, all classes can be scheduled to meet during m time periods.

Now we show that all classes cannot be scheduled in less than m hours. We prove this by contradiction. Suppose we can schedule the classes in k one-hour time periods, where $k < m$. We can then k-color the graph G_0 by coloring with the same color all vertices which correspond to classes meeting at the same time. To see that this is a legitimate k-coloring of G_0, consider two adjacent vertices. These vertices correspond to two classes containing one or more common students. Hence, these classes meet at different times, and thus the two vertices are colored differently. However, a k-coloring of G_0 produces a contradiction, since $\chi(G_0) = m$. ∎

Theorem 9.6 would completely solve our scheduling problem except for one unfortunate fact: ordinarily, it is extremely difficult to determine the chromatic number of a graph. There is no known formula that gives the chromatic number of a graph. However, we present the following result related to chromatic numbers. We write $\Delta(G)$ to denote the maximum degree among the vertices of G.

Theorem 9.7

For any graph G,

$$\chi(G) \leq 1 + \Delta(G).$$

Proof

We prove the result by induction on the order p of the graph. There is only one graph with $p = 1$, namely K_1. Since $\chi(K_1) = 1$ and $\Delta(G) = 0$,

$$\chi(K_1) \leq 1 + \Delta(K_1) = 1 + 0$$

so the result is true for $p = 1$.

Assume the result is true for all graphs with $p = k - 1$ vertices, and let G be a graph with $p = k$ vertices. We show that there exists a $(1 + \Delta(G))$-coloring of G, implying that $\chi(G) \leq 1 + \Delta(G)$. Let v be a vertex of G and consider the graph $G - v$. Since $G - v$ has $k - 1$ vertices, it follows by the induction hypothesis that $\chi(G - v) \leq 1 + \Delta(G - v)$. Hence there exists a $(1 + \Delta(G - v))$-coloring of $G - v$. Let such a coloring of $G - v$ be given. Now there are at most $\Delta(G)$ vertices adjacent to v in G, so no more than $\Delta(G)$ colors are used in coloring these vertices in $G - v$. If $\Delta(G - v) = \Delta(G)$, then some color used in coloring $G - v$ is available for v, and we have a $(1 + \Delta(G))$-coloring of G. If $\Delta(G - v) < \Delta(G)$, then we may introduce a new color for v, obtaining a coloring of G that requires no more than $1 + \Delta(G)$ colors. In any case, $\chi(G) \leq 1 + \Delta(G)$. ▮

Applying Theorem 9.7 to the graph H of Figure 9.11, we have $\chi(H) \leq 1 + \Delta(H) = 6$. However, we have already seen that $\chi(H) = 4$.

Problem Set 9.2

14. What is the chromatic number of K_p?

15. What is the chromatic number of a cycle?

16. What is the chromatic number of a tree?

17. What does Theorem 9.7 indicate about the chromatic number of $K(3, 3)$? $K(4, 4)$? $K(n, n)$? What is the chromatic number of $K(3, 3)$? $K(4, 4)$? $K(n, n)$?

*18. What can you say about the chromatic number of a bipartite graph? What do you think a natural definition of "tripartite graph" is? What can you say about the chromatic number of a tripartite graph?

*19. Prove that $\chi(G_1) = 3$ for the graph G_1 of Figure 9.10.

*20. Prove that $\chi(G_2) = 4$ for the graph G_2 of Figure 9.10.

21. A mathematics department chairman plans to offer seven graduate courses next semester, namely combinatorics (C), group theory (G), linear programming (L), numerical analysis (N), probability (P), statistics (S), and topology (T). The mathematics graduate students and the courses they plan to take are:

Android: C, L, T	Ginny: N, P
Bob: C, G, S	Homer: G, L
Carol: G, N	Inga: C, T
Don: C, L	Janet: C, S, T
Ed: L, N	Ken: P, S
Fred: C, G	Linda: P, T

How many time periods are needed for these seven courses?

*22. Let $U = \{1, 2, 3, \ldots, 10\}$. Furthermore, consider the following subsets of U:

$$\begin{aligned} A_1 &= \{1, 2, 6, 9\} & A_9 &= \{1, 3, 4, 5, 9\} \\ A_2 &= \{2, 7, 8\} & A_{10} &= \{1, 2, 3, 7\} \\ A_3 &= \{4, 6, 10\} & A_{11} &= \{2, 4, 6, 7\} \\ A_4 &= \{3, 5, 8\} & A_{12} &= \{1, 3, 5\} \\ A_5 &= \{6, 7, 9\} & A_{13} &= \{2, 4, 6, 10\} \\ A_6 &= \{1, 2, 3, 4, 6\} & A_{14} &= \{5, 8\} \\ A_7 &= \{8, 9\} & A_{15} &= \{7, 8, 9\} \\ A_8 &= \{1, 2, 6, 10\} \end{aligned}$$

Determine the minimum number n such that $S_1, S_2, \ldots, S_n$ are collections of the given subsets of U with the properties that each subset of U belongs to exactly one S_i, $1 \leq i \leq n$, and every two subsets of U belonging to the same S_i are disjoint.

9.3
The Four Color Problem

In the first two sections of this chapter we discussed planar graphs and chromatic numbers. In this section we mention a problem involving both concepts, and which is one of the most famous problems in all mathematics.

The Four Color Problem

TO DESCRIBE THIS PROBLEM, we begin with a map, divided into countries, say. We then assign a color to each country so that adjacent countries (i.e., countries sharing some common boundary) are assigned different colors. What is the least number of colors required to color all countries in the map? Of course, the answer depends on the countries and their geographic relationship to each other. Many mathematicians thought that no map, no matter how complicated, would require more than four colors. Whether this is true or not became known as the *Four Color Problem.*

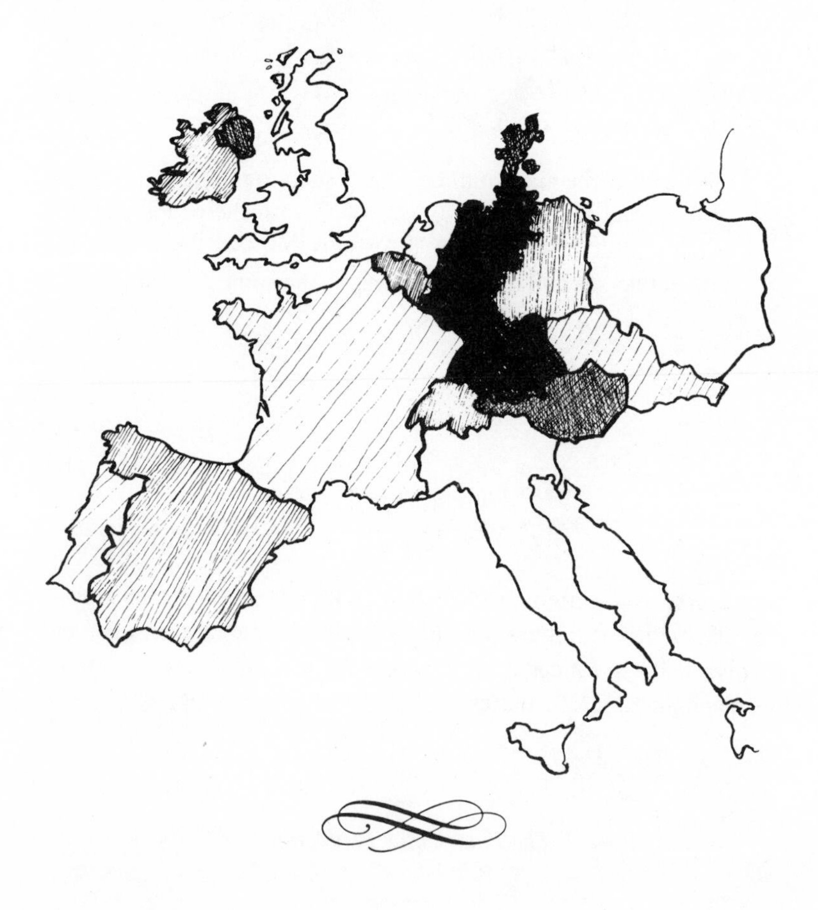

This famous problem originated in 1852, and was finally solved in 1976 when Appel and Haken showed it was true indeed that every map could be colored with four or fewer colors. It might seem that this should have ended interest in the Four Color Problem; however, such was not the case, due to the unusual nature of the solution. Appel and Haken solved the Four Color Problem by dividing the problem into nearly two thousand cases, according to the arrangements of countries within a map. To determine the possible ways of assigning colors in the various arrangements, they wrote computer programs to analyze the various colorings in each arrangement. After 1200 hours of computer calculations, they declared the problem solved!

Even though the solution of the Four Color Problem must be classified as a monumental achievement, many mathematicians were dissatisfied with (and some even skeptical of) the proof. Thus, a new problem arose; namely, does there exist a purely mathematical proof, unaided by computers, showing that every map can be colored with four or fewer colors? It is quite likely that no solution to *this* problem will be found during your lifetime (particularly if you are old).

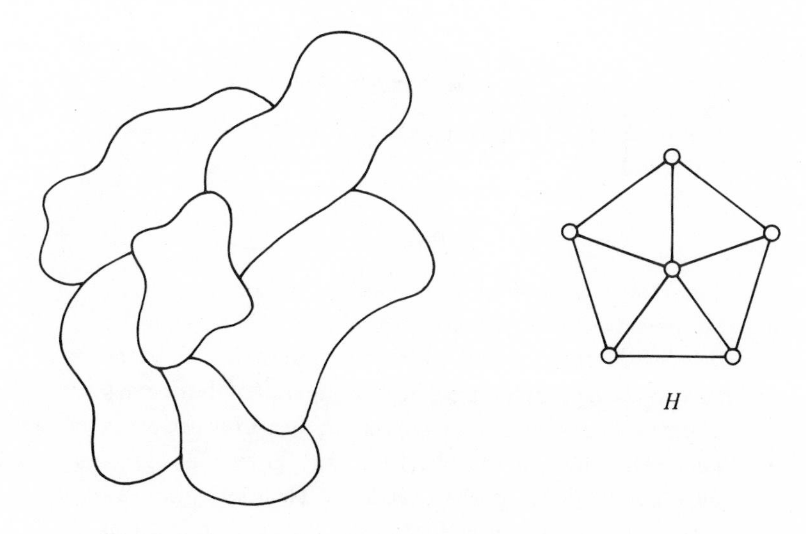

Figure 9.12

The Four Color Problem is closely related to graph theory. With each map we can associate a graph G whose vertices correspond to the countries, and where two vertices are adjacent if the corresponding countries are adjacent. Necessarily, each such graph G is a planar graph. Figure 9.12 shows a map and the corresponding planar graph H. We have already seen that $\chi(H) = 4$. Since every map can be colored with at most four colors, we have the following theorem.

The Four Color Theorem

If G is a planar graph, then $\chi(G) \leq 4$.

Despite the fact that it is extraordinarily difficult to prove the Four Color Theorem, it is, surprisingly, not particularly difficult to prove the Five Color Theorem. To prove this theorem, the following result is useful.

Theorem 9.8

Every planar graph G contains a vertex v such that $\deg v \leq 5$.

Proof

The result is obvious if G has six or fewer vertices. Suppose G is a (p, q) graph, where $p \geq 7$. If we sum the degrees of the vertices of G, we obtain $2q$. If every vertex has degree 6 or greater, then the sum of the degrees of the vertices of G is at least $6p$, i.e., $2q \geq 6p$. On the other hand, by Theorem 9.2 we have $q \leq 3p - 6$, so $2q \leq 6p - 12$. This is a contradiction. Therefore, not all vertices can have degree 6 or more, and so there exists a vertex v for which $\deg v \leq 5$. ▮

Theorem 9.9
The Five Color Theorem

If G is a planar graph, then $\chi(G) \leq 5$.

Proof

The proof is by induction on the order p of the graph. For $p = 1$, the result is obvious.

Assume that all planar graphs with $p = k - 1$ have chromatic number at most 5, and let G be a planar graph of order k. By Theorem 9.8, G contains a vertex v such that $\deg v \leq 5$. Let us draw G as a plane graph, and consider the plane graph $G - v$. Since $G - v$ has order $k - 1$, it follows by the induction hypothesis that $\chi(G - v) \leq 5$. This means we can perform a 5-coloring of $G - v$. We now assume that we have a 5-coloring of $G - v$, denoting the colors by 1, 2, 3, 4, and 5. If one of these colors is not used in coloring the vertices adjacent with v, then we may assign that color to v, producing a 5-coloring of G. Hence, we may assume that $\deg v = 5$ and all five colors are used for the vertices adjacent with v.

Suppose v_1, v_2, v_3, v_4, v_5 are the five vertices adjacent with v, arranged cyclically about v, and suppose v_1 has been colored 1, v_2 has been colored 2, etc. We now show that it is possible to recolor certain vertices of $G - v$, including a vertex adjacent with v, so a color becomes available for v. Consider the colors 1 and 3, and all the vertices of $G - v$ which have been colored 1 or 3. Of course, v_1 is colored 1 and v_3 is colored 3. In $G - v$ there may or may not be a v_1-v_3 path whose vertices are all colored 1 or 3 (see Figure 9.13). First we assume there is no such path. We consider all paths starting at v_1 whose vertices are all colored 1 or 3. These paths produce a subgraph of $G - v$, which we denote by H. Necessarily, v_3 is not in H; in fact, no vertex adjacent to v_3 is in H.

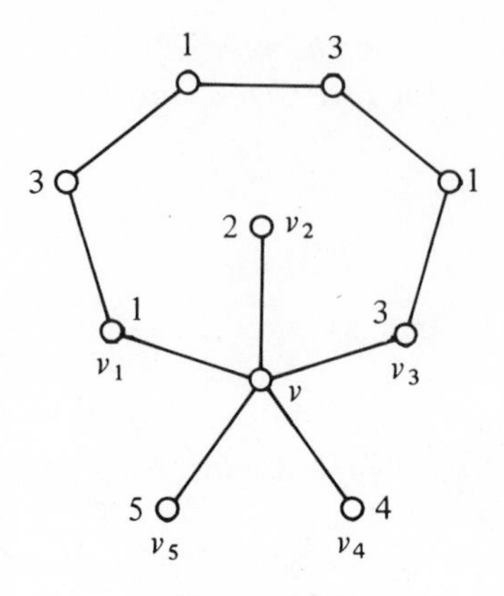

Figure 9.13

Now we interchange the colors of the vertices of H, producing another 5-coloring of $G - v$. However, this 5-coloring of $G - v$ assigns the color 3 to both v_1 and v_3. Thus, we may now assign the color 1 to vertex v, producing a 5-coloring of G. Therefore $\chi(G) \leq 5$.

Suppose, then, there does exist a v_1-v_3 path P in $G - v$, whose vertices are all colored 1 or 3. (If we interchanged the colors 1 and 3 of the vertices of H in this case, v_1 would be colored 3 and v_3 would be colored 1, and no color would be available for v. So we proceed in a different manner.) The path P together with the path v_3, v, v_1 produces a cycle in G which either encloses v_2 or encloses v_4 and v_5. Hence there does not exist a v_2-v_4 path in $G - v$ whose vertices are all colored 2 or 4. As before, we consider all paths beginning at v_2 whose vertices are colored 2 or 4. These paths produce a subgraph of $G - v$ which we represent by F. We interchange the colors of the vertices in F, producing a new 5-coloring of $G - v$ in which both v_2 and v_4 are colored 4. Hence, we may assign the color 2 to v, giving us a 5-coloring of G. Thus, $\chi(G) \leq 5$. ▮

Problem Set 9.3

23. What is the smallest order of a planar graph G for which $\chi(G) = 4$?

*24. Show that every planar graph of order at least two contains at least two vertices whose degrees are five or less.

*25. Show that it is impossible to improve Theorem 9.8 by replacing "5" by "4" in its statement; that is, show there exists a planar graph whose every vertex has degree 5 or more.

26. Show that $\chi(K_5) = 5$. Does this contradict the Four Color Theorem? Explain.

27. Does there exist a nonplanar graph with chromatic number one? two? three?

Suggestions for Further Reading

A great deal has been written about planar graphs. For the reader who wishes to learn more about the subject, the article by Harary [2] is a good place to begin.

The Four Color Problem has fascinated the mathematical world for well over a century. An interesting historical account of this problem prior to its solution was written by May [4]. Two other informative articles on the Four Color Problem are due to Harary [3] and Saaty [5]. A description of the proof of the Four Color Theorem by Appel and Haken may be found in [1].

[1] W. Haken, "An attempt to understand the Four Color Problem." *Journal of Graph Theory*, **1** (1977). To appear.

[2] F. Harary, "Recent results in topological graph theory." *Acta Mathematica Academiae Scientiarum Hungaricae*, **15** (1964), pp. 405–412.

[3] F. Harary, "The four color conjecture and other graphical diseases." *Proof Techniques in Graph Theory*. Academic Press, New York (1969), pp. 1–9.

[4] K. O. May, "The origin of the four-color conjecture." *Isis*, **56** (1965), pp. 346–348.

[5] T. L. Saaty, "Thirteen colorful variations on Guthrie's four-color conjecture." *American Mathematical Monthly*, **79** (1972), pp. 2–43.

*Chapter 10

Graphs and Other Mathematics

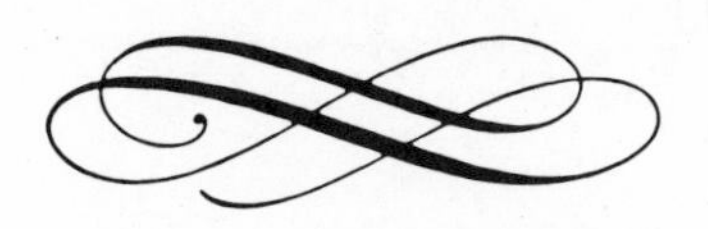

Graph theory has close relationships with several mathematical areas. In this chapter we consider three of these areas: matrices, topology, and groups.

10.1
Graphs and Matrices

A graph is completely determined by its vertex set and by a knowledge of which pairs of vertices are adjacent. This same information can easily be given by a matrix. In fact, much of graph theory could be investigated as a subject within matrix theory. There are certain advantages to this approach, since matrices can serve as computer input to work a variety of computations. On the other hand, there are several disadvantages to representing graphs as matrices, for this destroys the visual aspect of graphs, and the recognition of certain graphical properties in matrices is not necessarily simpler than in the diagram of a graph.

Let G be a graph of order p with vertices denoted by $v_1, v_2, \ldots, v_p$. Then the *adjacency matrix* $A = A(G) = [a_{ij}]$ is that $p \times p$ matrix in which $a_{ij} = 1$ if v_i and v_j are adjacent and $a_{ij} = 0$ otherwise. Thus, the matrix A is a (0, 1) matrix (i.e., every entry of A is 0 or 1); the main diagonal of A consists entirely of 0's (i.e., $a_{ii} = 0$ for $i = 1, 2, \ldots, p$); and A is symmetric (i.e., $a_{ij} = a_{ji}$ for $1 \leq i \leq p$ and $1 \leq j \leq p$). A graph G and its adjacency matrix are given in Figure 10.1.

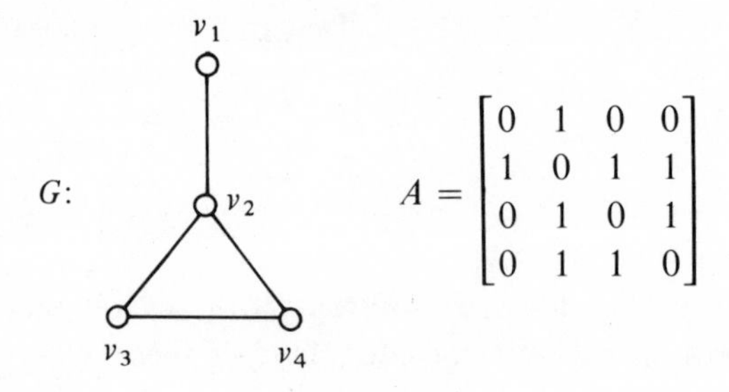

$$A = \begin{bmatrix} 0 & 1 & 0 & 0 \\ 1 & 0 & 1 & 1 \\ 0 & 1 & 0 & 1 \\ 0 & 1 & 1 & 0 \end{bmatrix}$$

Figure 10.1

Note that the adjacency matrix of a graph G ordinarily depends on how we label the vertices. For example, the graph G of Figure 10.1 is shown again in Figure 10.2 with a different labeling, resulting in a different adjacency matrix. Although the matrices of Figures 10.1 and 10.2 are unequal (as matrices), they represent isomorphic graphs.

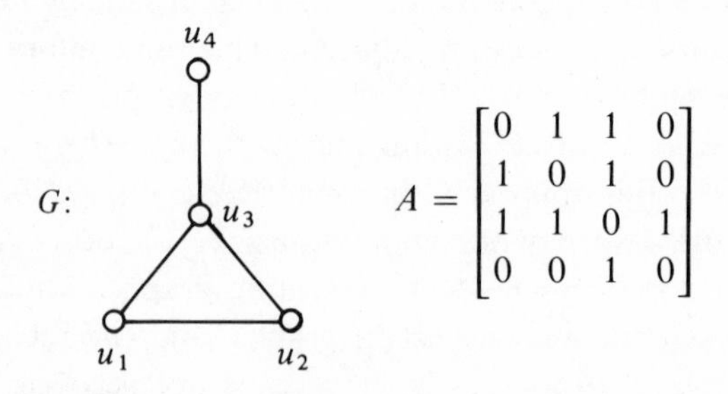

$$A = \begin{bmatrix} 0 & 1 & 1 & 0 \\ 1 & 0 & 1 & 0 \\ 1 & 1 & 0 & 1 \\ 0 & 0 & 1 & 0 \end{bmatrix}$$

Figure 10.2

One interesting property of an adjacency matrix of a graph concerns the entries of its various powers. The *length* of a walk is the number of occurrences of edges in it, and two u-v walks of the same length are *equal* if their edges occur in exactly the same order.

Theorem 10.1

Let A be the adjacency matrix of a graph G, where $V(G) = \{v_1, v_2, \ldots, v_p\}$. Then the (i, j) entry of A^n, $n \geq 1$, is the number of different v_i-v_j walks of length n in G.

Proof

The proof is by induction on n. For $n = 1$, we are considering the adjacency matrix A itself. The graph G contains a v_i-v_j walk of length one if and only if $v_i v_j$ is an edge of G. If the (i, j) entry is 1, then $v_i v_j$ is an edge of G; if the (i, j) entry is 0, then $v_i v_j$ is not an edge of G. Therefore, the result is true for $n = 1$.

Let $A^{n-1} = [a_{ij}^{(n-1)}]$, $n \geq 2$; that is, denote the (i, j) entry of the $(n - 1)$st power of A by $a_{ij}^{(n-1)}$. Assume $a_{ij}^{(n-1)}$ is the number of different v_i-v_j walks of length $n - 1$ in G. We now consider A^n, where, say, $A^n = [a_{ij}^{(n)}]$. Since $A^n = A^{n-1} \cdot A$, we can use the definition of matrix multiplication to obtain

$$a_{ij}^{(n)} = \sum_{k=1}^{p} a_{ik}^{(n-1)} a_{kj}.$$

Every v_i-v_j walk of length n in G consists of a v_i-v_k walk of length $n - 1$, where v_k is adjacent to v_j, followed by the edge $v_k v_j$ and the vertex v_j. Thus, $a_{ij}^{(n)}$ counts the number of different v_i-v_j walks of length n in G. ▮

G: v_1, v_2, v_3, v_4

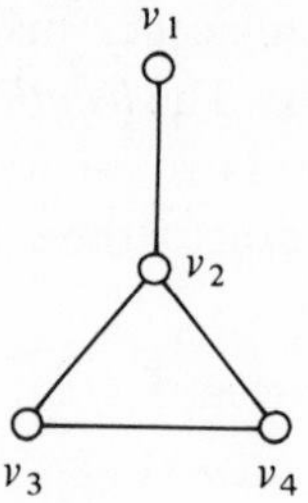

$$A = \begin{bmatrix} 0 & 1 & 0 & 0 \\ 1 & 0 & 1 & 1 \\ 0 & 1 & 0 & 1 \\ 0 & 1 & 1 & 0 \end{bmatrix}$$

$$A^2 = \begin{bmatrix} 1 & 0 & 1 & 1 \\ 0 & 3 & 1 & 1 \\ 1 & 1 & 2 & 1 \\ 1 & 1 & 1 & 2 \end{bmatrix} \qquad A^3 = \begin{bmatrix} 0 & 3 & 1 & 1 \\ 3 & 2 & 4 & 4 \\ 1 & 4 & 2 & 3 \\ 1 & 4 & 3 & 2 \end{bmatrix}$$

Figure 10.3

As an illustration of Theorem 10.1, we again turn to the graph G of Figure 10.1 and its adjacency matrix A. We compute A^2 and A^3 as in Figure 10.3. The (1, 2) entry of A^3 in Figure 10.3 is 3; hence, there are three v_1-v_2 walks of length 3 in G, namely

$$W_1: \ v_1, v_2, v_3, v_2; \qquad W_2: \ v_1, v_2, v_4, v_2;$$

$$W_3: \ v_1, v_2, v_1, v_2.$$

We now consider a second matrix which can be associated with a graph. Let G be a graph of order p and size q such that $V(G) = \{v_1, v_2, \ldots, v_p\}$ and $E(G) = \{e_1, e_2, \ldots, e_q\}$. The *incidence matrix* $B = B(G) = [b_{ij}]$ is that $p \times q$ matrix in which $b_{ij} = 1$ if v_i is incident with e_j, and $b_{ij} = 0$ otherwise. As with the adjacency matrix, the incidence matrix is a (0, 1) matrix. However, in general, the incidence matrix is not symmetric; in fact, in general, it is not a square matrix.

The particular incidence matrix obtained depends on how we label both vertices and edges. Figure 10.4 gives an example.

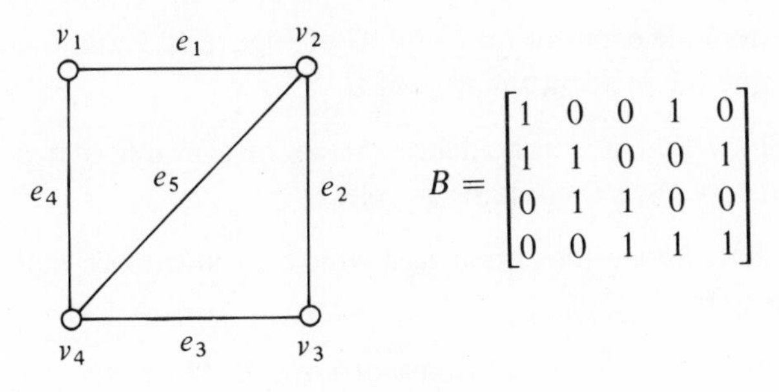

Figure 10.4

Problem Set 10.1

1. Label the vertices of the graph $G = K(2, 3)$ and determine the adjacency matrix $A = A(G)$. Compute the matrix A^2 and interpret its entries, using Theorem 10.1.

2. Let $G = K(n, n)$, $n \geq 2$. Furthermore, let $V(G) = V_1 \cup V_2$, where $V_1 = \{v_1, v_2, \ldots, v_n\}$ and $V_2 = \{v_{n+1}, v_{n+2}, \ldots, v_{2n}\}$, and $v_i v_j \in E(G)$ if and only if $v_i \in V_1$ and $v_j \in V_2$. Determine the adjacency matrix $A = A(G)$. Then, without performing any matrix multiplication, determine A^2 and A^3.

3. We pointed out that if A is the adjacency matrix of a graph G of order p, then A is a $p \times p$, (0, 1), symmetric matrix with 0's along the main diagonal. Show that these properties are sufficient to characterize adjacency matrices; that is, if A is a $p \times p$, (0, 1), symmetric matrix with 0's along the main diagonal, then A is the adjacency matrix of some graph G.

4. Determine the adjacency matrix A of the graph G of Figure 10.4. Without performing any matrix multiplication, determine A^2 and A^3.

5. If A is the adjacency matrix of a graph G and $A^2 = [a_{ij}^{(2)}]$, describe $a_{ij}^{(2)}$, $i \neq j$, and $a_{ii}^{(2)}$ without using the word "walk."

6. Label the edges of the graph G of Figure 10.1 and then determine the incidence matrix of G.

7. Show that if the incidence matrix of a graph G is a square matrix, then G contains a cycle.

8. Characterize those matrices which are incidence matrices of graphs.

9. Let M^t denote the transpose of a matrix M, i.e., the first row of M^t is the first column of M, the second row of M^t is the second column of M, etc. For the graph G of Figure 10.4, compute the product $B \cdot B^t$. If B is the incidence matrix of a graph G, try to describe what you can expect, in general, for the product $B \cdot B^t$.

10. Define another matrix which can be associated with a graph.

10.2
Graphs and Topology

Some aspects of graph theory are very closely tied to the mathematical field of topology, particularly that topic called "surface topology." Actually, we discussed relationships between graphs and surfaces in Chapter 9 when we considered planar graphs and the Four Color Problem.

As we already noted, we can draw only certain graphs in the plane so that no edges intersect except at vertices; these are called planar graphs. Such a "drawing" is also called an *embedding* of the graph in the plane. It is not too difficult to see that planar graphs are precisely those graphs which can be embedded on the surface of a sphere. For example, Figure 10.5 shows the graph K_4 embedded on a sphere. Hence, "spherical graphs" and planar graphs are exactly the same. Of course, every graph can be embedded in 3-dimensional space.

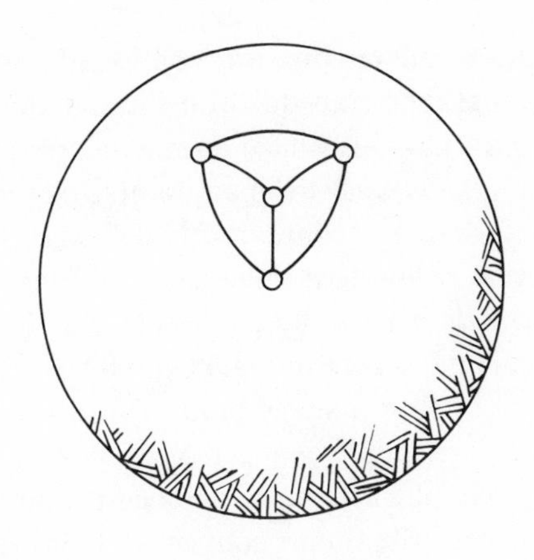

Figure 10.5

Another kind of surface that plays an important role in topology is the doughnut-shaped surface called the torus. Figure 10.6 shows the graph K_4 embedded on a torus. If G is a graph embedded on the torus, then the *regions* of G are the connected pieces of the torus remaining after we remove the vertices and edges of G. (Recall the definition of "region" in the case of plane graphs.) In the case of the graph K_4 embedded on the torus in Figure 10.6, there are four regions.

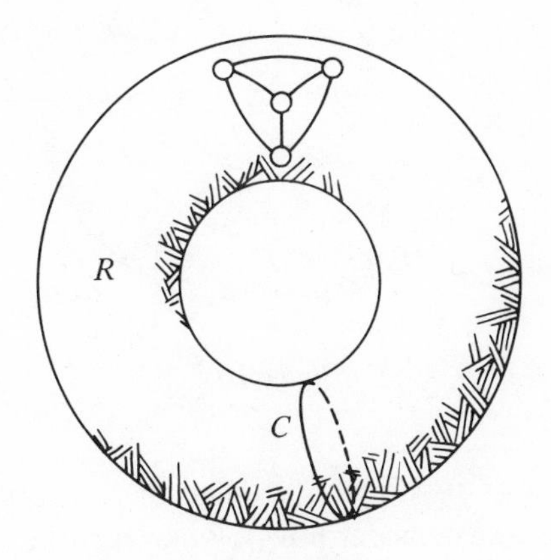

Figure 10.6

By definition, all regions are connected, whether they are determined from graphs embedded on the sphere or on the torus. However, a region may also have another important property. A region is called *simply connected* if any simple closed curve (such as a circle or ellipse) can be "continuously deformed" (i.e., shrunk or contracted) in that region to a single point. For example, in Figure 10.6 the region R is *not* simply connected, for if we consider the simple closed curve C shown in region R, then C *cannot* be continuously deformed in R to a single point. All three other regions of K_4 in Figure 10.6 are simply connected.

If a connected (planar) graph is embedded on the sphere, then every region is necessarily simply connected. However, this need not be the case for connected graphs embedded on the torus. A graph is called *toroidal* if it can be embedded on the torus. Every planar graph is toroidal; however, the converse statement is false, i.e., there are nonplanar graphs which can be embedded on the torus. One such example is K_5, shown embedded on the torus in Figure 10.7.

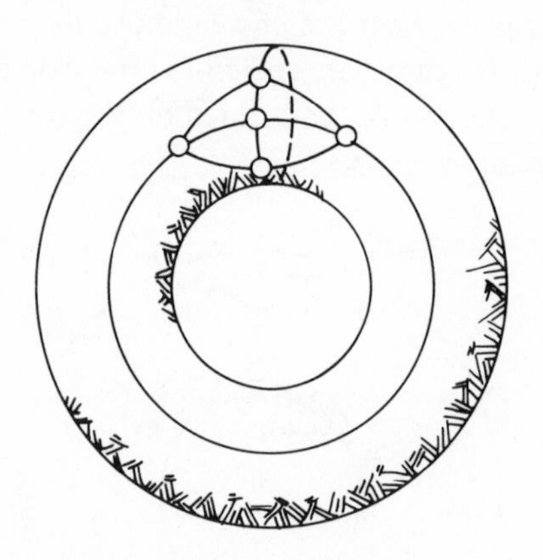

Figure 10.7

Another way of drawing graphs on the torus often proves convenient. First we consider how to construct the torus. We begin with a rectangle, and then we bend it in the shape of a cylinder so that the top and bottom sides meet. We then bend it again so that the two ends meet, thereby obtaining a torus. Figure 10.8 illustrates this process.

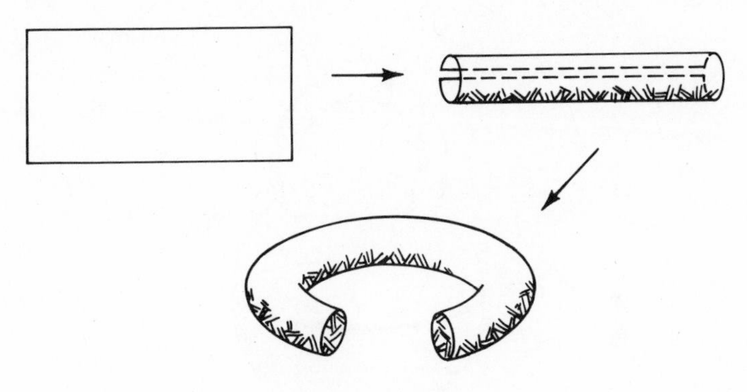

Figure 10.8

Hence, in the original rectangle of Figure 10.8, opposite sides have been identified with each other. Therefore, the points labeled A in Figure 10.9(a) are the same, as are the points labeled B. This identification of opposite sides is denoted by the arrows going in the same direction. Thus, the graph K_5 can be embedded on the torus, as shown in Figure 10.9(b).

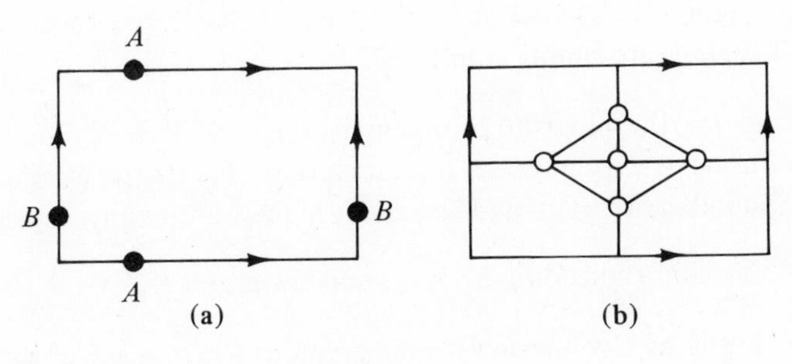

Figure 10.9

In topology, a torus is also known as a sphere with one "handle," as shown in Figure 10.10. We say that the torus and the sphere with one handle are "topologically equivalent," or that they are "homeomorphic."

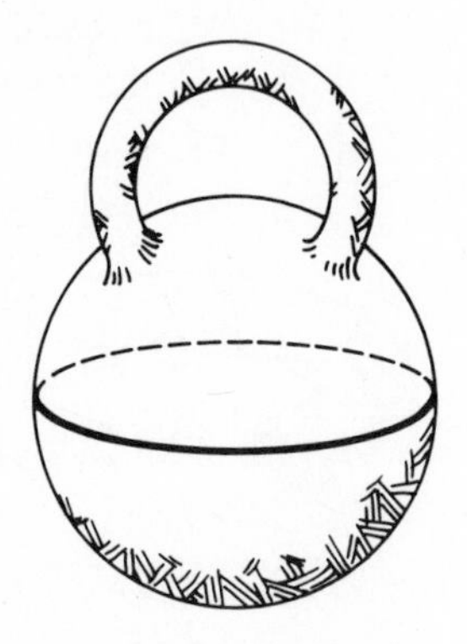

Figure 10.10

Problem Set 10.2

11. Figure 10.7 shows K_5 embedded on the torus. How many regions are simply connected?

12. Draw the bipartite (nonplanar) graph $K(3, 3)$ on the torus using its rectangular representation. Try to determine the number of regions and how many of these are simply connected.

13. Give an embedding of the nonplanar graph K_6 on the torus.

14. Let G be a connected planar graph. Is the number of regions determined by an embedding of G on the sphere always

equal to the number of regions determined by an embedding of G on the torus? Explain.

15. Theorem 9.1 states that if G is a connected graph embedded in the plane and G has p vertices, q edges, and r regions, then the number $p - q + r$ always equals 2. Another theorem states that if G is a connected graph embedded on the torus so that *each region is simply connected*, and G has p vertices, q edges, and r regions, then $p - q + r$ always equals the same number. Try to determine the value of this number.

16. What do you think a "double torus" is? Give an embedding of K_7 on the double torus. (It turns out that it is possible to embed K_7 even on the single torus, but it is not possible to embed K_8 on the torus.)

17. Show that it is possible to embed a given graph G on a sphere with n handles if we make n large enough.

18. Learn what a Möbius strip is. Explain why every planar graph can be embedded on the Möbius strip and give an embedding of K_5 on the Möbius strip.

10.3
Graphs and Groups

First, let us present (or recall) the definition of a group. By a binary operation $\otimes$ on a nonempty set Γ, we mean a function associating an element of Γ with each ordered pair (a, b) of (not necessarily distinct) elements of Γ. We denote this element of Γ by $a \otimes b$. We emphasize the word "ordered" in the definition, as it need *not* be the case that $a \otimes b = b \otimes a$.

A *group* is a nonempty set Γ together with a binary operation $\otimes$ (usually called multiplication) on Γ satisfying the following three properties:

(1) *Associative Law*: For any elements a, b, c (not necessarily distinct) of Γ,

$$a \otimes (b \otimes c) = (a \otimes b) \otimes c.$$

(2) There exists an element e of Γ (called the *identity* of Γ) such that

$$a \otimes e = e \otimes a = a$$

for every element a of Γ.

(3) For each element a of Γ, there exists an element a^{-1} of Γ (called the *inverse* of a) such that

$$a \otimes a^{-1} = a^{-1} \otimes a = e.$$

Although groups can take a variety of forms, many important groups arise from permutations (see Section A.4 in the Appendix). We consider some special examples of these groups.

Let G be a graph of order p with vertices labeled $v_1, v_2, \ldots, v_p$, say. It is obvious that G is isomorphic to itself; hence there exists an isomorphism α from G to G, i.e., a one-to-one mapping $\alpha: V(G) \to V(G)$ which preserves adjacency. In fact, the obvious mapping is $\alpha(v_i) = v_i$ for $i = 1, 2, \ldots, p$. However, there may well be other mappings (isomorphisms) in addition to this obvious one.

An isomorphism from a graph to itself is called an *automorphism* of the graph. The set of all automorphisms of a graph G under the binary operation of composition (again refer to Section A.4 of the Appendix) is always a group, called the *automorphism group*, or simply *the group* of G, and is denoted by $\mathscr{A}(G)$. Before proceeding further, let us consider an illustration.

Example 10.1

Let G be the graph of Figure 10.11, with vertices labeled 1, 2, 3; that is, $V(G) = \{1, 2, 3\}$. There are six permutations

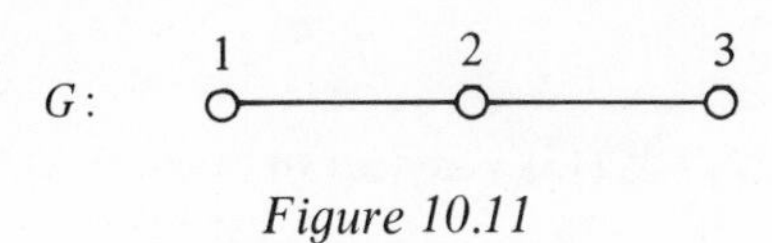

Figure 10.11

on $V(G)$; however, only two of these are isomorphisms (automorphisms) of G. The obvious automorphism is

$$\alpha_0 = \begin{pmatrix} 1 & 2 & 3 \\ 1 & 2 & 3 \end{pmatrix};$$

however,

$$\alpha_1 = \begin{pmatrix} 1 & 2 & 3 \\ 3 & 2 & 1 \end{pmatrix}$$

is also an automorphism. Let us consider the composition $\alpha_1 \circ \alpha_0$. [Recall that the composition $f \circ g$ of functions f and g means that $(f \circ g)(a) = f(g(a))$ for each element a on which g is defined.] We see that

$$(\alpha_1 \circ \alpha_0)(1) = \alpha_1(\alpha_0(1)) = \alpha_1(1) = 3,$$
$$(\alpha_1 \circ \alpha_0)(2) = \alpha_1(\alpha_0(2)) = \alpha_1(2) = 2,$$

and

$$(\alpha_1 \circ \alpha_0)(3) = \alpha_1(\alpha_0(3)) = \alpha_1(3) = 1.$$

Hence,

$$\alpha_1 \circ \alpha_0 = \begin{pmatrix} 1 & 2 & 3 \\ 3 & 2 & 1 \end{pmatrix} = \alpha_1.$$

We also investigate $\alpha_1 \circ \alpha_1$. Here we have

$$(\alpha_1 \circ \alpha_1)(1) = \alpha_1(\alpha_1(1)) = \alpha_1(3) = 1,$$
$$(\alpha_1 \circ \alpha_1)(2) = \alpha_1(\alpha_1(2)) = \alpha_1(2) = 2,$$

and

$$(\alpha_1 \circ \alpha_1)(3) = \alpha_1(\alpha_1(3)) = \alpha_1(1) = 3.$$

Therefore,

$$\alpha_1 \circ \alpha_1 = \begin{pmatrix} 1 & 2 & 3 \\ 1 & 2 & 3 \end{pmatrix} = \alpha_0.$$

For the graph G of Figure 10.11 we have $\mathscr{A}(G) = \{\alpha_0, \alpha_1\}$. Since $\mathscr{A}(G)$ has two elements, this group has "order" two. The graph G is displayed again in Figure 10.12 together with its group, including the "group table."

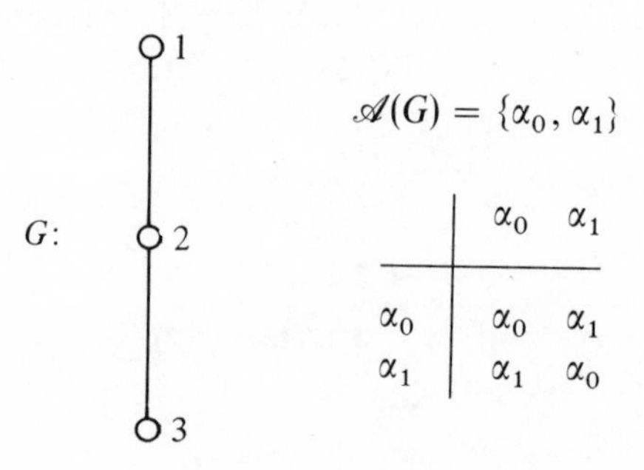

	α_0	α_1
α_0	α_0	α_1
α_1	α_1	α_0

Figure 10.12

With the aid of the group table in Figure 10.12, we can now see that the identity element of $\mathscr{A}(G)$ is α_0, the inverse of α_0 is α_0, and the inverse of α_1 is α_1.

Example 10.2

Consider the cycle C_4 of Figure 10.13. In addition to the "identity" automorphism

$$\alpha_0 = \begin{pmatrix} 1 & 2 & 3 & 4 \\ 1 & 2 & 3 & 4 \end{pmatrix},$$

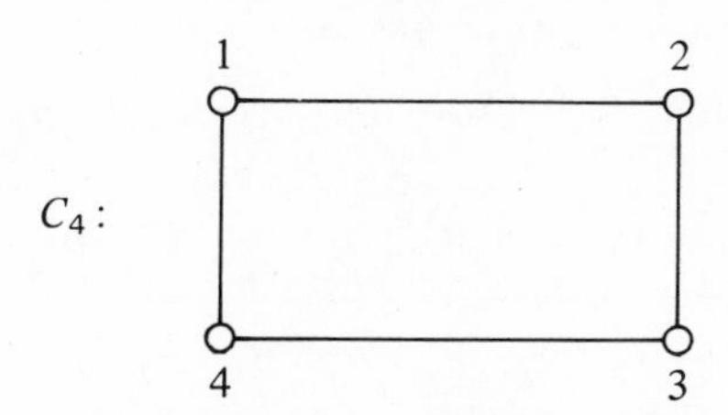

Figure 10.13

there are three automorphisms which might be considered "rotations" of the cycle C_4:

$$\alpha_1 = \begin{pmatrix} 1 & 2 & 3 & 4 \\ 2 & 3 & 4 & 1 \end{pmatrix}, \qquad \alpha_2 = \begin{pmatrix} 1 & 2 & 3 & 4 \\ 3 & 4 & 1 & 2 \end{pmatrix},$$

$$\alpha_3 = \begin{pmatrix} 1 & 2 & 3 & 4 \\ 4 & 1 & 2 & 3 \end{pmatrix}.$$

The group $\mathscr{A}(C_4)$ also contains four other automorphisms which we can interpret as "reflections" of the cycle C_4:

$$\beta_1 = \begin{pmatrix} 1 & 2 & 3 & 4 \\ 4 & 3 & 2 & 1 \end{pmatrix}, \qquad \beta_2 = \begin{pmatrix} 1 & 2 & 3 & 4 \\ 2 & 1 & 4 & 3 \end{pmatrix},$$

$$\beta_3 = \begin{pmatrix} 1 & 2 & 3 & 4 \\ 3 & 2 & 1 & 4 \end{pmatrix}, \qquad \beta_4 = \begin{pmatrix} 1 & 2 & 3 & 4 \\ 1 & 4 & 3 & 2 \end{pmatrix}.$$

The group $\mathscr{A}(C_4)$ is summarized in Figure 10.14. We refer to the groups of the cycles C_n as *dihedral groups*, and denote them by D_n. Hence, Figure 10.14 shows the dihedral group D_4 (of order 8).

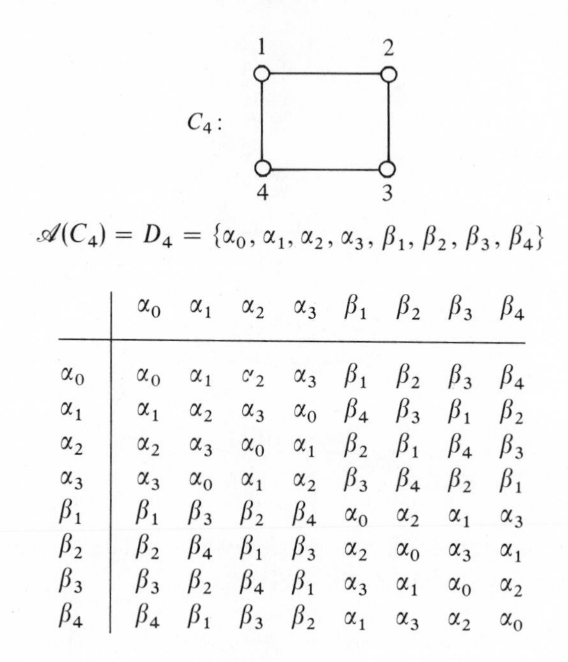

$$\mathscr{A}(C_4) = D_4 = \{\alpha_0, \alpha_1, \alpha_2, \alpha_3, \beta_1, \beta_2, \beta_3, \beta_4\}$$

	α_0	α_1	α_2	α_3	β_1	β_2	β_3	β_4
α_0	α_0	α_1	α_2	α_3	β_1	β_2	β_3	β_4
α_1	α_1	α_2	α_3	α_0	β_4	β_3	β_1	β_2
α_2	α_2	α_3	α_0	α_1	β_2	β_1	β_4	β_3
α_3	α_3	α_0	α_1	α_2	β_3	β_4	β_2	β_1
β_1	β_1	β_3	β_2	β_4	α_0	α_2	α_1	α_3
β_2	β_2	β_4	β_1	β_3	α_2	α_0	α_3	α_1
β_3	β_3	β_2	β_4	β_1	α_3	α_1	α_0	α_2
β_4	β_4	β_1	β_3	β_2	α_1	α_3	α_2	α_0

Figure 10.14

We have seen that we can associate a group with every graph G. We consider the reverse question now of associating a graph with a given group. In this context, we consider only groups of finite order. A group Γ is *generated* by the elements $g_1, g_2, \ldots, g_k$ (and these elements are called *generators*) if every element of Γ can be expressed as the product of a finite number of generators and their inverses. A group with one generator is called a *cyclic group*.

Let Γ be a given finite group with $\Delta = \{g_1, g_2, \ldots, g_k\}$ a set of generators for Γ. It is common not to include the identity in a generating set. We describe a graph (actually a digraph) called the *Cayley color graph* of Γ, denoted by $D_\Delta(\Gamma)$. (The symbol "Δ" is included in the notation since the Cayley color graph ordinarily depends on the set of generators chosen. In general, several Cayley color graphs may exist for a given group, depending on the generating set.) We associate a vertex of $D_\Delta(\Gamma)$ with each element of group Γ. With each generator g_i of Γ we associate a color, say color i. Suppose $\Gamma = \{x_1, x_2, \ldots, x_p\}$ and vertex v_i of $D_\Delta(\Gamma)$ corresponds to x_i for $i = 1, 2, \ldots, p$. We draw an arc colored i from v_1 to v_2 if and only if $x_2 = x_1 g_i$. This gives us the Cayley color graph of Γ.

Example 10.3

Consider the group $\Gamma = \{e, a, b, c, d\}$ with the following group table:

	e	a	b	c	d
e	e	a	b	c	d
a	a	b	c	d	e
b	b	c	d	e	a
c	c	d	e	a	b
d	d	e	a	b	c

The group Γ is cyclic with generator set $\Delta = \{a\}$ because

$$a = a, \quad b = a^2, \quad c = a^3, \quad d = a^4, \quad \text{and} \quad e = a^5.$$

For convenience we label the vertices of the Cayley color graph $D_\Delta(\Gamma)$ in Figure 10.15 with the elements of

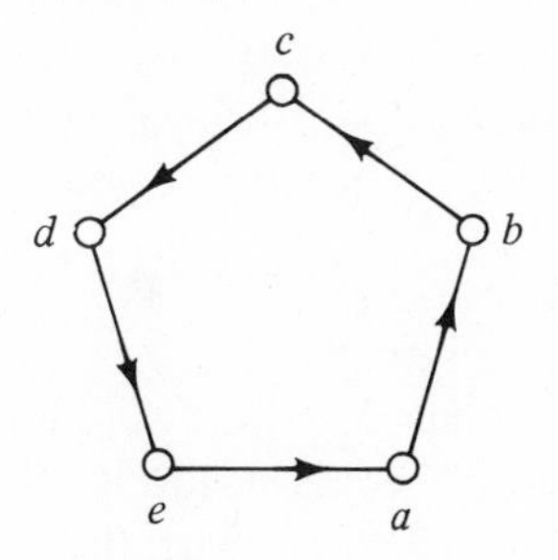

Figure 10.15

the group. In this case all arcs of the same color. For example, there is an arc from b to c because $c = ba$, as can be seen in the group table. Again, other Cayley color graphs can be obtained for $\Gamma = \{a, b, c, d, e\}$ by selecting different generating sets, such as $\Delta' = \{a, b\}$, for example.

Example 10.4

Next we consider the group Γ of all permutations on the set $\{1, 2, 3\}$. Here, $\Gamma = \{x_1, x_2, x_3, x_4, x_5, x_6\}$, where

$$x_1 = \begin{pmatrix} 1 & 2 & 3 \\ 1 & 2 & 3 \end{pmatrix}, \qquad x_2 = \begin{pmatrix} 1 & 2 & 3 \\ 2 & 3 & 1 \end{pmatrix},$$

$$x_3 = \begin{pmatrix} 1 & 2 & 3 \\ 3 & 1 & 2 \end{pmatrix}, \qquad x_4 = \begin{pmatrix} 1 & 2 & 3 \\ 1 & 3 & 2 \end{pmatrix},$$

$$x_5 = \begin{pmatrix} 1 & 2 & 3 \\ 3 & 2 & 1 \end{pmatrix}, \qquad x_6 = \begin{pmatrix} 1 & 2 & 3 \\ 2 & 1 & 3 \end{pmatrix}.$$

The group table in this case is

	x_1	x_2	x_3	x_4	x_5	x_6
x_1	x_1	x_2	x_3	x_4	x_5	x_6
x_2	x_2	x_3	x_1	x_6	x_4	x_5
x_3	x_3	x_1	x_2	x_5	x_6	x_4
x_4	x_4	x_5	x_6	x_1	x_2	x_3
x_5	x_5	x_6	x_4	x_3	x_1	x_2
x_6	x_6	x_4	x_5	x_2	x_3	x_1.

Here we observe that $\Delta = \{x_2, x_6\}$, for example, is a generator set for Γ. We associate a solid arc with the color of x_2 and a dashed arc with the color of x_6. This Cayley color graph is shown in Figure 10.16.

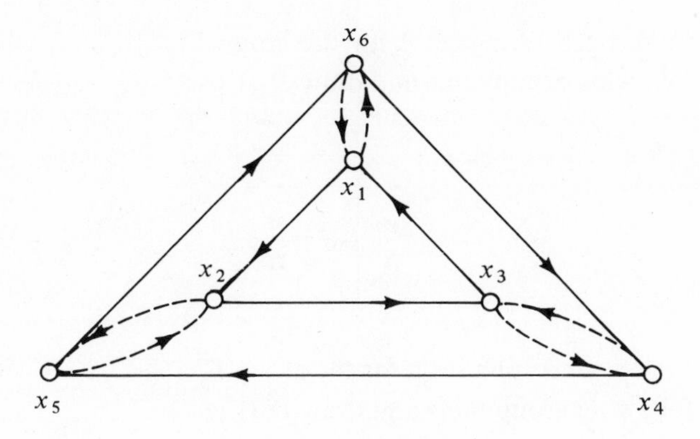

Figure 10.16

Problem Set 10.3

19. Determine $\mathscr{A}(K_2)$.
20. Determine $\mathscr{A}(\overline{K}_2)$.
21. Determine $\mathscr{A}(K(1, 3))$.
22. Determine $\mathscr{A}(K_3)$.
23. Determine $\mathscr{A}(P_n)$, where P_n is a path with $n \geq 2$ vertices.
24. The dihedral group D_n has order $2n$. Describe the $2n$ automorphisms of C_n, the cycle of order n and size n.
25. Show that for any graph G, the groups $\mathscr{A}(G)$ and $\mathscr{A}(\overline{G})$ are the same (isomorphic).
26. Find a generating set Δ for the group $\Gamma = \{e, a\}$ with the following group table and draw $D_\Delta(\Gamma)$:

	e	a
e	e	a
a	a	e

27. Find a generating set Δ for the group $\Gamma = \{e, a, b\}$ with the following group table and draw $D_\Delta(\Gamma)$:

	e	a	b
e	e	a	b
a	a	b	e
b	b	e	a

28. Find a generating set Δ for the group $\Gamma = \{e, a, b, c\}$ with the following group table and draw $D_\Delta(\Gamma)$:

	e	a	b	c
e	e	a	b	c
a	a	e	c	b
b	b	c	e	a
c	c	b	a	e

29. Draw $D_\Delta(\mathscr{A}(K_2))$.

30. The graph $G = C_4$ and $\mathscr{A}(C_4) = \mathscr{A}(G)$ are given in Figure 10.14. Find a generating set Δ for the group $\mathscr{A}(G)$ and draw $D_\Delta(\mathscr{A}(G))$.

Suggestions for Further Reading

Harary [3] has written a very readable expository article on graphs and matrices. Furthermore, the books by Harary [2] and Behzad, Chartrand, and Lesniak-Foster [1] each contains a chapter on the subject.

A survey of "topological graph theory" can be found in the article by Kainen [4], while the book by White [5] discusses relationships among "graphs, groups, and surfaces."

[1] M. Behzad, G. Chartrand, and L. Lesniak-Foster, *Graphs & Digraphs*. Prindle, Weber & Schmidt, Boston. To be published.

[2] F. Harary, *Graph Theory*. Addison-Wesley, Reading, Mass. (1969).

[3] F. Harary, "Graphs and matrices." *SIAM Review*, **9** (1967), pp. 83–90.

[4] P. C. Kainen, "Some recent results in topological graph theory." *Graphs and Combinatorics* (R. A. Bari and F. Harary, eds.), Springer-Verlag, Berlin (1974), pp. 76–108.

[5] A. T. White, *Graphs, Groups, and Surfaces*. North-Holland, Amsterdam (1973).

Appendix

Sets, Relations, Functions, and Proofs

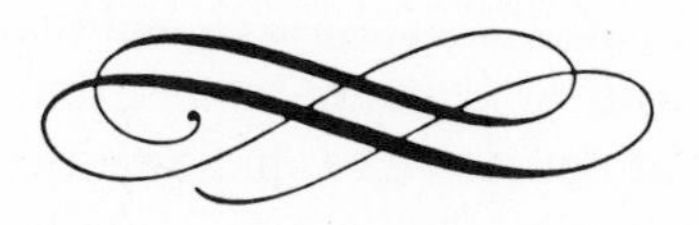

Every area of mathematics makes use to some extent of the concepts "set," "relation," and "function." Graph theory not only involves these concepts, but, in fact, "set" and "relation" are fundamental to the meaning of the word "graph" itself. It is therefore essential that we clearly understand these terms. As with every other mathematical subject, the theory of graphs develops through theorems and proofs. We also include a brief discussion of this important topic.

A.1
Sets and Subsets

We give no technical definition of the concept of *set*, for any attempt to define "set" only uses words whose meaning is synonymous with "set" itself, such as "collection" or "family." We say that a set is composed of *elements*, but we will not attempt to define this term either. We will soon clarify how these words are employed.

To describe a set, we list its elements, if this is feasible, or we may present some rule indicating precisely those elements belonging to the set. For example, we may refer to the set of students taking this class. If the number of students is not very large, we can describe this same set by listing the individuals; however, a listing of the students does not suggest what these people have in common, namely that they are all members of this class. Hence we may prefer to describe a set by means of a rule (if possible), independent of the number of elements in the set.

For the most part, we shall use capital (upper case) letters to denote a set, such as U or V, and lower case letters to represent the elements of a set. The notation $u \in U$ indicates that u is an element of U, while $v_1, v_2 \in V$ implies that v_1 and v_2 are both elements of the set V. If w is not an element of W, we write $w \notin W$. To indicate that a set U consists of the elements u_1, u_2, and u_3, we write $U = \{u_1, u_2, u_3\}$. If N denotes the natural numbers (positive integers), then

$$V = \{v_i | 1 \leq i \leq 100, i \in N\}$$

means that V consists of the 100 elements v_1, v_2, etc., up to and including v_{100}. More formally, this is read as "the set of all elements v_i such that [the vertical bar means "such that"] $1 \leq i \leq 100$ and $i \in N$." Indeed, this same set V could be described equally well by writing

$$V = \{v_1, v_2, \ldots, v_{100}\},$$

$$V = \{v_i | i = 1, 2, \ldots, 100\},$$

or

$$V = \{v_i | 1 \leq i \leq 100\},$$

since it is understood that $i \in N$.

If U and V are sets with the property that every element of U is an element of V, then we say U is a *subset* of V, and write $U \subseteq V$. If U and V contain precisely the same elements, then U and V are *equal sets*, denoted $U = V$. The statement $U = V$ is equivalent to saying $U \subseteq V$ and $V \subseteq U$.

In accord with the preceding statements,

$$\{u_1, u_2, u_2\} = \{u_1, u_2\} = \{u_2, u_1\};$$

that is, the composition of a set is not affected by an element listed more than once, nor by the order in which elements are listed. If $U \subseteq V$ and $U \neq V$ (that is, U is not equal to V), then every element of U is an element of V, but at least one element of V does not belong to U. Under these conditions, we say U is a *proper* subset of V, and write $U \subset V$. A set may have no elements; if this is the case, we designate the set by $\varnothing$ and call it the *null set* or *empty set*. For every set W, it follows that $\varnothing \subseteq W$; for if this were not the case, there would exist an element x such that $x \in \varnothing$ and $x \notin W$. However, this is impossible since $\varnothing$ contains no elements.

In any particular discussion concerning sets, the sets involved are usually subsets of some specified set, called the *universal set*. For example, at a certain time we may be interested in sets of positive integers; in this case, the universal set is N. If U denotes the universal set on some occasion and S is a subset of U, then the *complement* $\bar{S}$ of S is that set consisting of all elements of U not belonging to S, i.e.,

$$\bar{S} = \{x \in U \mid x \notin S\}.$$

Note that $\bar{U} = \varnothing$.

Given two sets V and W, the *union* of V and W, denoted $V \cup W$, is the set consisting of elements belonging to V or W (or both). The *intersection* $V \cap W$ of V and W is the set containing those elements belonging to both V and W. If $V \cap W = \varnothing$, then we say V and W are *disjoint sets* (meaning that V and W have no elements in common). As an illustration, let

$$U = \{u_1, u_2, u_3\}, \quad V = \{u_1, u_3\}, \quad \text{and} \quad W = \{u_2, u_4\}.$$

Then

$$U \cup V = \{u_1, u_2, u_3\}, \quad U \cap V = \{u_1, u_3\},$$

$$V \cup W = \{u_1, u_2, u_3, u_4\}, \quad U \cap W = \{u_2\}, \quad \text{and} \quad V \cap W = \varnothing.$$

Therefore, V and W are disjoint.

Two famous set equalities involving complements, unions, and intersections are called *DeMorgan's Laws*, stated next. We prove the first of these laws and leave the proof of the second as an exercise.

Theorem A.1
(DeMorgan's Laws)

Let U be the universal set, and let A and B be subsets of U. Then:

(*a*) $\overline{A \cup B} = \bar{A} \cap \bar{B}$

(*b*) $\overline{A \cap B} = \bar{A} \cup \bar{B}$.

Proof of (*a*)

We give the proof in two parts; specifically, we show (i) $\overline{A \cup B} \subseteq \bar{A} \cap \bar{B}$ and (ii) $\bar{A} \cap \bar{B} \subseteq \overline{A \cup B}$.

To prove (i), we let $x \in \overline{A \cup B}$. Then $x \in U$ but $x \notin A \cup B$. Since $x \notin A \cup B$, the element x belongs to neither A nor B, i.e., $x \notin A$ and $x \notin B$. This implies that $x \in \bar{A}$ and $x \in \bar{B}$. Therefore, $x \in \bar{A} \cap \bar{B}$. We have thus shown that $\overline{A \cup B} \subseteq \bar{A} \cap \bar{B}$.

To prove (ii), we let $y \in \bar{A} \cap \bar{B}$. Hence, $y \in \bar{A}$ and $y \in \bar{B}$, so that $y \notin A$ and $y \notin B$. Since y belongs to neither A nor B, it follows that $y \notin A \cup B$, implying that $y \in \overline{A \cup B}$. This verifies (ii). ▮

Problem Set A.1

1. How many elements are contained in the set

$$U = \{u_i | 1 < i < 10, i \in N\}?$$

2. Let U, V, and W be nonempty sets, and suppose that for every element $w \in W$, either $w \in U$ or $w \in V$. Show that $W \subseteq U \cup V$. (*Hint*: Let x be any element of W and show that $x \in U \cup V$.)

3. Let V and W be disjoint sets. Suppose $V_1 \subseteq V$ and $W_1 \subseteq W$. Show that V_1 and W_1 are disjoint sets. (*Hint*: Either V_1 and W_1 are disjoint sets or they are not. Suppose they are not. Then there is an element $x \in V_1 \cap W_1$. Show that this is impossible.)

4. Let $V_1 = \{v_1, v_3, v_4\}$, $V_2 = \{v_2, v_5\}$, $V_3 = \{v_1, v_3\}$, and $V_4 = \{v_3, v_4\}$. Determine the union and the intersection of all pairs of these sets. Which sets are disjoint?

5. Let U, V, and W be sets such that $U \subset V$ and $V \subseteq W$. Prove that $U \subset W$.

6. Prove part (b) of Theorem A.1.

A.2
Cartesian Products and Relations

For elements u_1 and u_2 of some set, we write (u_1, u_2) to denote the *ordered pair* of these two elements. Here, the order in which we write the elements is important; u_1 is the first element and u_2 is the second element. Hence the ordered pair (u_2, u_1) is not the same as (u_1, u_2). This differs from the meaning of $\{u_1, u_2\}$, where the fact that u_1 is written first has no special significance; that is, $\{u_1, u_2\} = \{u_2, u_1\}$.

The *Cartesian product* $V \times W$ of two sets V and W is the set of all ordered pairs such that the first element of each pair belongs to V and the second element of each pair belongs to W; in symbols,

$$V \times W = \{(v, w) \mid v \in V, w \in W\}.$$

If $V = \{v_1, v_2\}$ and $W = \{w_1, w_2, w_3\}$, then

$$V \times W = \{(v_1, w_1), (v_1, w_2), (v_1, w_3), (v_2, w_1), (v_2, w_2), (v_2, w_3)\};$$

also

$$V \times V = \{(v_1, v_1), (v_1, v_2), (v_2, v_1), (v_2, v_2)\}.$$

Let $|S|$ denote the number of elements in a set S. (The number $|S|$ is called the *cardinality* of S.) For a given element v in a set V, the number of ordered pairs in $V \times W$ having v as the first element is $|W|$. Since there are $|V|$ different first elements among the ordered pairs being considered, it follows that the total number of ordered pairs in $V \times W$ is $|V| \cdot |W|$, i.e., $|V \times W| = |V| \cdot |W|$.

By a *relation* on a set V, we mean any subset of $V \times V$. Thus, if $|V| = n$ for some finite set V, then a relation on V may have as many as n^2 elements, or as few as zero elements. For example, suppose $V_0 = \{a, b, c\}$. Then one possible relation on V_0 is

$$R = \{(a, b), (b, b), (b, c), (c, b)\}.$$

We say that *a is related to b* by R because $(a, b) \in R$. However, b is not related to a in this case since $(b, a) \notin R$. We can also indicate that $(a, b) \in R$ by writing $a\, R\, b$. We can then write $b \not{R}\, a$.

There are several interesting properties of relations on sets. A relation R on a set V is *reflexive* if for every $x \in V$, the ordered pair (x, x) belongs to R. If $V_0 = \{a, b, c\}$, then any reflexive relation on V_0 must include the ordered pairs (a, a), (b, b), and (c, c). A relation R on a set V is *irreflexive* if for every $x \in V$, it follows that $(x, x) \notin R$. We emphasize that if a relation is not reflexive, it need not be irreflexive. For example, if $V_1 = \{y, z\}$, then $S = \{(y, y), (y, z)\}$ is a relation on V_1 which is neither reflexive nor irreflexive.

A relation R on a set V is *symmetric* if whenever $(x, y) \in R$, then $(y, x) \in R$; the relation is *asymmetric* if whenever $(x, y) \in R$, where $x \neq y$, then $(y, x) \notin R$. A relation R on V is called *transitive* if whenever $(x, y) \in R$ and $(y, z) \in R$, then $(x, z) \in R$. Thus, if a relation R on some set V is not symmetric, there exists some ordered pair $(a, b) \in R$, where $a \neq b$, such that $(b, a) \notin R$. If a relation R is not transitive, then there are ordered pairs (a, b) and (b, c) in R such that $(a, c) \notin R$.

Example A.1

Let $V = \{u, v, w, x\}$ and $E = \{(u, v), (v, u), (v, w), (w, v)\}$. Then E is a relation on V which is irreflexive, symmetric, and not transitive.

Example A.2

Suppose V_0 denotes the set of all living people. Let R_1 denote the relation in which $(a, b) \in R_1$ if $a, b \in V_0$ and a is the same age (in years) as b. Then R_1 is reflexive, symmetric, and transitive. Let R_2 denote the relation in which $(a, b) \in R_2$ if $a, b \in V_0$ and a is *not* the same age as b. If $(a, a) \in R_2$, then this implies that the person a is not the same age as himself or herself, which, of course, is impossible. Hence, for every $a \in V_0$, we have $(a, a) \notin R_2$, and so R_2 is irreflexive. If $(a, b) \in R_2$, i.e., if a is not the same age as b, then necessarily b is not the same age as a. Therefore, $(b, a) \in R_2$ and R_2 is symmetric. Finally, suppose $(a, b) \in R_2$ and $(b, c) \in R_2$. Then a is not the same age as b and b is not the same age as c; but we cannot conclude that $(a, c) \in R_2$. For example, a may be 19 years old, b 21 years old, and c 19 years old. In this case, $(a, c) \notin R_2$. Hence, R_2 is not transitive.

Problem Set A.2

7. Which of the following relations have the properties reflexive, symmetric, transitive on the indicated sets?
   (a) "Is parallel to" on the set of lines in the plane.
   (b) "Is less than 100 miles from" on the set of cities of the world.
   (c) "Is the sister of" on the set of all living people.

8. Let $V = \{a, b, c, d\}$, and let $R = \{(a, a), (a, b), (a, c), (a, d), (b, b), (b, c), (b, d), (c, c), (c, d), (d, d)\}$. Which properties (reflexive, irreflexive, symmetric, asymmetric, transitive) does the relation R possess?

9. Let $V = \{a, b, c, d\}$. Give an example of a relation on V which is not reflexive, symmetric, or transitive.

10. Let $V = \{a, b, c, d\}$. Determine a reflexive, symmetric, transitive relation on V containing (a, b), (b, c), and (c, d). Show that there is only one relation having these properties.

11. Let V be a nonempty set, and let $R = \varnothing$ be the "empty" relation on V. Does R have any of the reflexive, symmetric, and transitive properties?

12. Let $V = \{a, b, c\}$. Give an example of a relation on V which is:
    (a) reflexive and symmetric, but not transitive.
    (b) reflexive and transitive, but not symmetric.
    (c) symmetric and transitive, but not reflexive.

A.3
Equivalence Relations

Relations which are reflexive, symmetric, and transitive are called equivalence relations and play an important role in all branches of mathematics. More formally, a relation R on a set A is called an *equivalence relation* if the following three properties are satisfied:

(i) $a\,R\,a$ for all $a \in A$;

(ii) $a\,R\,b$ implies $b\,R\,a$ for all $a, b \in A$;

(iii) $a\,R\,b$ and $b\,R\,c$ implies $a\,R\,c$ for all $a, b, c \in A$.

Examples of equivalence relations are numerous. Rather than presenting examples dealing with abstract sets, we illustrate equivalence relations on sets which we can describe in a less abstract manner. All of the following relations are equivalence relations on the indicated sets:

(1) "Is equal to" on the set of integers. (This follows since $a = a$ for every integer a; if $a = b$ then $b = a$; and if $a = b$ and $b = c$, then $a = c$.)

(2) "When divided by the positive integer m, has the same remainder as" on the set of integers.

(3) "Is parallel to or coincides with" on the set of lines in the plane.

(4) "Is congruent to" on the set of triangles in the plane.

(5) "Has the same first name as" on the set of all people with first names.

(6) "Has the same number of pages as" on the set of all books.

One of the most important properties of equivalence relations deals with what are called "partitions" of sets. We now consider this property.

Let R be an equivalence relation on a set A. Then for $a \in A$, we define the subset $[a]$ as follows:

$$[a] = \{x \mid x \in A, x\, R\, a\}.$$

The sets $[a]$, $a \in A$, are called *equivalence sets* or *equivalence classes.* Since R is an equivalence relation, R is reflexive, and therefore $a \in [a]$.

If $A_1, A_2, \ldots, A_n$ are nonempty, pairwise disjoint subsets (i.e., every two subsets are disjoint) of a nonempty set A such that the union of all subsets is A, then the set $\{A_1, A_2, \ldots, A_n\}$ of these subsets is called a *partition* of A. With these definitions in mind, we can now present the result we are interested in.

Theorem A.2

Let R be an equivalence relation on a nonempty set A. Then the set $S = \{[a] | a \in A\}$ of equivalence classes is a partition of the set A.

Proof

We must show that the set of equivalence classes satisfies the definition of a partition of A. First we observe that if $[c]$ is an equivalence class, say, then $[c]$ is nonempty since $c \in [c]$. Furthermore, for an arbitrary element $b \in A$, we have $b \in [b]$; therefore, the union of the elements of S is A.

It remains for us to show that every two distinct elements of S are disjoint. Suppose $[b]$ and $[c]$ are two distinct elements of S. We shall assume $[b]$ and $[c]$ are not disjoint; then, under this assumption, we shall prove $[b] = [c]$. This, however, contradicts the fact that $[b]$ and $[c]$ are distinct, implying that our assumption that $[b]$ and $[c]$ are not disjoint is incorrect. Therefore, $[b]$ and $[c]$ *are* disjoint.

Since we are assuming $[b]$ and $[c]$ are not disjoint, it follows that $[b] \cap [c] \neq \varnothing$. This implies that there exists an element $x \in A$ such that $x \in [b] \cap [c]$; hence, $x \in [b]$ and $x \in [c]$. Thus, $x \, R \, b$ and $x \, R \, c$. Since R is symmetric, $b \, R \, x$, and since R is transitive, $b \, R \, c$ (so that $c \, R \, b$).

We now show that $[b] = [c]$. We do this by verifying that each of $[b]$ and $[c]$ is a subset of the other. Suppose y is any element of $[b]$, so $y \, R \, b$. Because $b \, R \, c$ and R is transitive, we conclude that $y \, R \, c$ and $y \in [c]$; that is, any element of $[b]$ is also an element of $[c]$. Therefore, $[b] \subseteq [c]$. Now suppose $z \in [c]$, so $z \, R \, c$. Since $c \, R \, b$, we have $z\,R\,b$ and $z \in [b]$. Thus, $[c] \subseteq [b]$, thereby proving $[b] = [c]$. This gives us the desired contradiction and completes the proof that the elements of S give a partition of A. ▮

Let us illustrate Theorem A.2. Let $A = \{1, 2, 3, 4, 5, 6\}$ and let $R = \{(1, 1), (1, 2), (1, 3), (2, 1), (2, 2), (2, 3), (3, 1), (3, 2), (3, 3), (4, 4), (4, 5), (5, 4), (5, 5), (6, 6)\}$. We can show that R is an equivalence relation on A. Hence, by Theorem A.2, $\{[a] \mid a \in A\}$ is a partition of A. In this case, the choices for a are 1, 2, 3, 4, 5, and 6. We have $[1] = \{1, 2, 3\}$, $[2] = \{1, 2, 3\}$, $[3] = \{1, 2, 3\}$, $[4] = \{4, 5\}$, $[5] = \{4, 5\}$ and $[6] = \{6\}$. Hence

$$\{[a] \mid a \in A\} = \{[1], [2], [3], [4], [5], [6]\} = \{[1], [4], [6]\},$$

the last equality following because $[1] = [2] = [3]$ and $[4] = [5]$. We now observe that $\{[1], [4], [6]\}$ is indeed a partition of A. This is shown pictorially in Figure A.1, where we see how the set A is partitioned by the equivalence relation R.

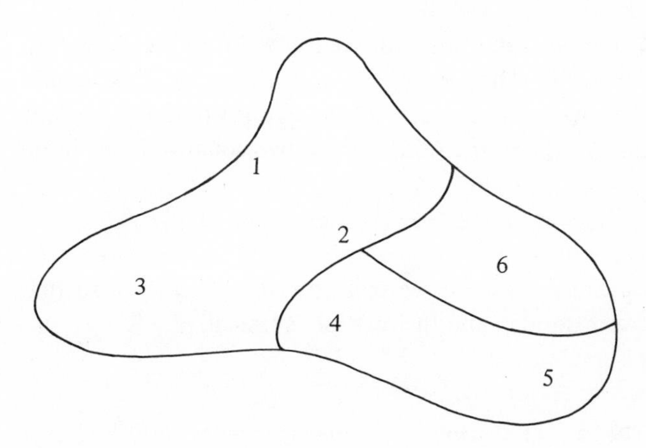

Figure A.1

We give one final (but important) example of an equivalence relation. Let Z denote the set of integers, and let $m, d \in Z$, where $d \neq 0$. We say *d divides m*, written $d|m$, if an integer q exists such that $m = dq$. Next, let $a, b, n \in Z$, where $n \geq 2$. We say *a is congruent to b modulo n* if $n|(a - b)$, and write $a \equiv b \pmod{n}$.

Theorem A.3

Let $n\,(\geq 2)$ be an integer. The relation "is congruent to modulo n" is an equivalence relation on Z.

Proof

Let $a \in Z$. Then $n|(a - a)$ since $a - a = n \cdot 0$. Therefore, $a \equiv a \pmod{n}$ for every $a \in Z$ and the relation is reflexive.

Let $a, b \in Z$ and suppose $a \equiv b \pmod{n}$. Hence $n|(a - b)$, which implies that an integer q exists such that $a - b = nq$. Thus, $b - a = n(-q)$. Since $-q \in Z$, it follows that $n|(b - a)$, so $b \equiv a \pmod{n}$ and the relation is symmetric.

Let a, b, $c \in Z$ and suppose $a \equiv b \pmod{n}$ and $b \equiv c \pmod{n}$. Thus, $n|(a - b)$ and $n|(b - c)$. This implies that there exist integers q_1 and q_2 such that $a - b = nq_1$ and $b - c = nq_2$. Adding these two equalities, we obtain

$$a - c = nq_1 + nq_2 = n(q_1 + q_2).$$

Since $q_1 + q_2 \in Z$, we conclude that $n|(a - c)$, so that $a \equiv c \pmod{n}$ and the relation is transitive. ∎

The set of distinct equivalence classes resulting from the equivalence relation of Theorem A.3 is referred to as the *integers modulo n* and is denoted by Z_n or $Z/(n)$. For example, if $n = 5$, then $Z_5 = \{[0], [1], [2], [3], [4]\}$, where

$$[0] = \{0, \pm 5, \pm 10, \pm 15, \ldots\},$$
$$[1] = \{1, -4, 6, -9, 11, \ldots\},$$
$$[2] = \{2, -3, 7, -8, 12, \ldots\},$$
$$[3] = \{3, -2, 8, -7, 13, \ldots\},$$

and

$$[4] = \{4, -1, 9, -6, 14, \ldots\}.$$

Problem Set A.3

13. Let A denote the set of all people, and let R denote the relation "is born in the same year as" on the set A. Show that R is an equivalence relation on A. Describe the elements of the set $\{[a] \mid a \in A\}$.

14. Let $A = \{1, 2, 3, 4, 5, 6\}$ and suppose R is an equivalence relation on A. If $1\,R\,3$, $3\,R\,4$, $2\,R\,6$, and $|\{[a] \mid a \in A\}| = 3$, determine the elements of R. Also, determine the distinct equivalence classes in this case.

15. Let A be a nonempty set, and let R be a relation on A with the property that $a\,R\,b$ for any elements a and b in A (a and b need not be distinct). Show that R is an equivalence relation on A. Determine the equivalence classes.

16. Let A be a nonempty set, and let $S = \{A_1, A_2, \ldots, A_n\}$ be a set of subsets of A forming a partition of A. Determine an equivalence relation R on A such that the set $\{[a] \mid a \in A\}$ is precisely the set S.

17. Describe the contents of the distinct elements of Z_2 and Z_3.

18. How many distinct elements are in Z_n, where $n \geq 2$? Describe the set Z_n.

19. Let R be a relation defined on Z by $a\,R\,b$ if $a + b \equiv 0 \pmod 2$. Show that R is an equivalence relation and determine the distinct equivalence classes.

20. Let R be a relation defined on Z by $a\,R\,b$ if $a + b \equiv 0 \pmod 3$. Show that R is *not* an equivalence relation.

21. Let R be a relation defined on Z by $a\,R\,b$ if $a^2 \equiv b^2 \pmod 5$. Show that R is an equivalence relation and determine the distinct equivalence classes.

A.4
Functions

Let A and B be two nonempty sets. Then a *function* f from A to B (denoted $f : A \to B$) is a set of ordered pairs (a, b), with $a \in A$ and $b \in B$, satisfying the property that for every element $a \in A$ there exists exactly one ordered pair in f containing a as the first element of the pair. For example, let A denote the set of all circles and let B denote the set of real numbers. We can define a function $f_1 : A \to B$ by letting $(a, b) \in f_1$ provided a is a circle and b is the area of a. With the same two sets A and B, we can define another function $f_2 : A \to B$ by letting $(a, b) \in f_2$ provided a is a circle and b is the circumference of a. Hence, the function f_1 associates an area with a circle and f_2 associates a circumference with a circle.

Let $A = \{a, b, c\}$ and $B = \{1, 2, 3, 4\}$. Then $g_1 = \{(a, 2), (b, 4)\}$ and $g_2 = \{(a, 3), (b, 1), (c, 2), (c, 4)\}$ are *not* functions from A to B, while $g_3 = \{(a, 4), (b, 4), (c, 2)\}$ *is* a function from A to B. Note that if A is a nonempty finite set with, say, n elements and B is any nonempty set, then every function $f : A \to B$ has exactly n ordered pairs.

If $f : A \to B$ is a function from A to B, then it is also customary to call f a *mapping* from A to B and say that f *maps* A into B. Furthermore, if $(a, b) \in f$, then we can write $b = f(a)$ and say a is mapped into

b by f and b is the *image* of a under f. According to the definition of "function," every element of A has exactly one image; on the other hand, an element of B may be the image of one, several, or no elements of A.

If a function $f : A \to B$ has the property that every element of B is the image of at least one element of A, then we say f is a mapping from A *onto* B. If $A = \{a, b, c\}$ and $B = \{1, 2, 3\}$, then $f = \{(a, 2), (b, 1), (c, 1)\}$ is a function from A into B, but f is *not* a function from A onto B since 3 is not the image of any element of A.

A function $f : A \to B$ is *one-to-one* if distinct elements of A have distinct images in B. Symbolically, f is one-to-one if $a_1, a_2 \in A$ and $a_1 \neq a_2$ imply that $f(a_1) \neq f(a_2)$. It is also logically equivalent to define a function $f : A \to B$ to be one-to-one if whenever $f(a_1) = f(a_2)$, then $a_1 = a_2$. If $A = \{a, b, c, d\}$ and $B = \{1, 2, 3, 4, 5\}$, then the function $f = \{(a, 1), (b, 4), (c, 5), (d, 4)\}$ is not one-to-one, since b and d have the same image, i.e., $f(b) = f(d)$.

Let $A = \{a_1, a_2, a_3, a_4\}$ and $B = \{b_1, b_2, b_3, b_4\}$. Then $f = \{(a_1, b_3), (a_2, b_2), (a_3, b_4), (a_4, b_1)\}$ is a one-to-one function from A onto B. You might already have observed that if f is a one-to-one function from a set A onto a set B, then f describes a pairing of the elements of A with the elements of B. This is always the case. Thus, a one-to-one mapping from a set A onto a set B is possible only when A and B have the same number of elements.

Of the many methods of combining two functions into a single function, the most important method is composition of functions. If $f : A \to B$ and $g : B \to C$, then the *composition* of f and g, denoted $g \circ f : A \to C$, is defined by $(g \circ f)(a) = g(f(a))$ for every $a \in A$. As an illustration, we let

$$A = \{a_1, a_2, a_3\}, \qquad B = \{b_1, b_2, b_3, b_4\}, \qquad C = \{c_1, c_2, c_3\},$$

$$f = \{(a_1, b_3), (a_2, b_1), (a_3, b_1)\},$$

and

$$g = \{(b_1, c_2), (b_2, c_3), (b_3, c_3), (b_4, c_1)\}.$$

Then

$$g \circ f = \{(a_1, c_3), (a_2, c_2), (a_3, c_2)\}.$$

The following theorems give two of the most important results dealing with composition.

Theorem A.4

Let f be a function from A onto B and let g be a function from B onto C. Then $g \circ f$ is a function from A onto C.

Proof

We must show that every element of C is the image of at least one element of A under the function $g \circ f$. Let $c_0 \in C$. Since g is a function from B onto C, there exists an element $b_0 \in B$ such that $g(b_0) = c_0$. Also, since f is a function from A onto B, there is an element $a_0 \in A$ such that $f(a_0) = b_0$. We now claim that c_0 is the image of a_0 under the composite function $g \circ f$. To see this, we need only observe that

$$(g \circ f)(a_0) = g(f(a_0)) = g(b_0) = c_0. \quad \blacksquare$$

Theorem A.5

Let f be a one-to-one function from A into B, and let g be a one-to-one function from B into C. Then $g \circ f$ is a one-to-one function from A into C.

Proof

We use the alternate, but equivalent, definition of one-to-one function. Therefore, let $a_1, a_2 \in A$ and suppose $(g \circ f)(a_1) = (g \circ f)(a_2)$. We must verify that $a_1 = a_2$.

From the given equation, we have $g(f(a_1)) = g(f(a_2))$, and since g is a one-to-one function, it follows that $f(a_1) = f(a_2)$. Now we may apply the hypothesis that f is one-to-one and arrive at the desired conclusion that $a_1 = a_2$. ∎

If A is a nonempty set and $f : A \to A$ is a one-to-one function which is also onto, then f is called a *permutation* of A. For example, if $A = \{a_1, a_2, a_3, a_4\}$ and $f = \{(a_1, a_3), (a_2, a_4), (a_3, a_1), (a_4, a_2)\}$, then f is a permutation of A. Note that a permutation of a set is a rearrangement of the elements of a set. The above function f rearranges (or permutes) the elements of the above set A as follows:

$$\begin{array}{lllll} \text{original order:} & a_1 & a_2 & a_3 & a_4 \\ \text{rearrangement:} & a_3 & a_4 & a_1 & a_2. \end{array}$$

This permutation is one of 24 permutations that exist on the set $A = \{a_1, a_2, a_3, a_4\}$.

Problem Set A.4

22. Give an example of two sets A_1 and B_1 and a function $f_1 : A_1 \to B_1$ such that f_1 is onto but not one-to-one.
23. Give an example of two sets A_2 and B_2 and a function $f_2 : A_2 \to B_2$ such that f_2 is one-to-one but not onto.
24. Determine all permutations of the set $V = \{v_1, v_2, v_3, v_4\}$.
25. Let f be a one-to-one function from a set A onto a set B. Define the relation $f^{-1} = \{(b, a) | (a, b) \in f\}$. Show that f^{-1} is a one-to-one function from B onto A. (f^{-1} is called the *inverse* of f.)

26. Let S be the set of all nonempty finite sets. Let $A_1, A_2 \in S$ and define a relation R on S by $(A_1, A_2) \in R$ if there exists a one-to-one mapping from A_1 onto A_2. Show that R is an equivalence relation on S. Describe the equivalence classes.

A.5
Theorems and Proofs

We have seen a few examples of theorems and proofs in this Appendix; the text itself contains many more. One of the chief difficulties in learning mathematics is understanding proofs; however, this is ordinarily not nearly as difficult as inventing proofs of your own. Unfortunately, inventing proofs is something which cannot be taught, in general. Constructing proofs is commonly learned through experience, by first trying to understand and then imitating the proofs of others. This, hopefully, leads to the creation of original proofs.

Certainly, you have little hope of proving a theorem if you do not have a sound understanding of the statement of the theorem. All theorems can be stated in the form of an *implication* "If P, then Q," where P and Q denote statements which are true or false, but not both. The statement P is the *hypothesis* of the theorem and Q is the *conclusion*. Implications are special kinds of statements, and therefore are true or false themselves. Theorems are true implications which are usually of special interest.

We often attempt to prove the implication "If P, then Q" by assuming that P is true and then trying to verify that Q is true. If we accomplish this, we have a *direct proof* of the implication. It is also logically correct to assume that Q is false and then show that P is false. This is an *indirect proof* of the implication. As a variation of this method, we may assume that P is true *and* that Q is false (hoping, of course, that this is impossible) and then attempt to show that P

is false. Since P cannot be both true and false, we have proved the implication by *contradiction.*

Example A.3

We illustrate these various proof techniques with a simple example. We begin by remarking that every integer is even or odd. If x is an even integer, then $x = 2n$ for some integer n. For example, 6 and -10 are even integers since they can be expressed as twice an integer; in particular, $6 = 2 \cdot (3)$ and $-10 = 2 \cdot (-5)$. An odd integer y can be expressed as $y = 2m + 1$, where m is an integer. For odd integers -5 and 17, we have $-5 = 2(-3) + 1$ and $17 = 2(8) + 1$. We can also say that x is even if $x \equiv 0 \pmod 2$ and y is odd if $y \equiv 1 \pmod 2$. With these remarks in mind, we are now prepared to give three proofs of the following elementary result.

Theorem

Let x be an integer. If x is even, then $y = x + 5$ is odd.

Direct Proof

Assume x is an even integer. Then $x = 2n$ for some integer n. So we have

$$y = x + 5 = 2n + 5 = (2n + 4) + 1 = 2(n + 2) + 1.$$

Since $m = n + 2$ is an integer, it follows that $y = 2m + 1$, where m is an integer. Therefore, y is an odd integer. ▮

Indirect Proof

Assume $y = x + 5$ is not an odd integer. Then y must be even. Therefore $y = x + 5 = 2n$, where n is an integer.

Hence,

$$x = 2n - 5 = (2n - 6) + 1 = 2(n - 3) + 1.$$

Since $m = n - 3$ is an integer, we have $x = 2m + 1$ for some integer m. Thus, x is odd; that is, x is not even. ▮

Proof by Contradiction

Assume x is an even integer. Then $x = 2n$ for some integer n. Suppose that $y = x + 5$ is not odd; thus, y is even so that $y = x + 5 = 2m$ for some integer m. Therefore,

$$x = 2m - 5 = (2m - 6) + 1 = 2(m - 3) + 1.$$

Since $k = m - 3$ is an integer, it follows that $x = 2k + 1$ for some integer k, so x is odd. This is a contradiction. ▮

We next consider some of the common terminology used in the statements of theorems. The theorem

"If P, then Q"

may also be stated

"P *only if* Q"

as well as

"P is *sufficient* for Q"

or "Q is *necessary* for P."

It is often very difficult to see that these four statements all say the same thing. In order to get a little better feeling for this, let us take a statement and rephrase it in three ways. Consider the statement

"*If* I get all A's this semester, *then* I will get an A in this course."

This statement may be rephrased as follows:

> "I will get all A's this semester *only if* I get an A in this course."
>
> "For me to get an A in this course, it is *sufficient* for me to get all A's this semester."
>
> "For me to get all A's this semester, it is *necessary* for me to get an A in this course."

The *converse* of "If P, then Q" is the implication "If Q, then P." The truth or falseness of "If P, then Q" does not imply the truth or falseness of its converse. From what we have said, the converse "If Q, then P" may also be stated as "Q only if P"; "Q is sufficient for P"; or "P is necessary for Q." Hence, if the implication "If P, then Q" and its converse "If Q, then P" (or equivalently, "P if Q") are both true, then we may state these two implications as "P if and only if Q" and "P is necessary and sufficient for Q."

We illustrate some of these terms with an example from geometry.

Example A.4

The implication "If a triangle has three equal sides, then it has two equal sides" is true. (It is highly doubtful that this implication would be classified as a theorem.) We may also state this implication as "A triangle has three equal sides only if it has two equal sides." Using the words "sufficient" and "necessary," we may say "A sufficient condition for a triangle to have two equal sides is that it have three equal sides" or "A necessary condition for a triangle to have three equal sides is that it have two equal sides." The converse of the implication is "If a triangle has two equal sides, then it has three equal sides." The converse is false.

Problem Set A.5

27. Prove the implication "If x is an odd integer, then $y = x - 3$ is an even integer" using the three proof techniques: direct proof, indirect proof, proof by contradiction.

28. (a) Give an indirect proof of the following result: "Let x and y be integers. If x is even, then $x + 2y$ is even."
    (b) Is the converse of the statement in (a) true or false?

29. Give a proof of the following result: "If $x = 3$, then $x^2 = 9$." What type of proof did you use?

30. State the implication "If $x = 3$, then $x^2 = 9$" using the phrases (a) only if; (b) sufficient; and (c) necessary.

31. Give an example of a nonmathematical implication of the form "If P, then Q," where P and Q are statements. Then rewrite the implication using the phrases (a) only if; (b) sufficient; and (c) necessary.

32. The implication "If a triangle has three equal sides, then it has three equal angles" and its converse are both true. State the implication and its converse as a single statement using the phrases (a) "if and only if"; and (b) "necessary and sufficient."

A.6
Mathematical Induction

One of the most important and most useful proof techniques is referred to as mathematical induction. Before describing this technique, we discuss some properties of integers.

It is a property of the set N of positive integers that they are *well-ordered*, i.e., every nonempty subset S of N has a smallest (or least) element. For example, the least element of N itself is the integer 1. For the integer m, we introduce the notation

$$N_m = \{x \mid x \in Z, x \geq m\}.$$

The set N_m is also well-ordered. Using this fact, we establish the following result.

Theorem A.6

Let S be a subset of N_m with the properties:

(1) $m \in S$;

(2) *If $x \in S$, then $x + 1 \in S$.*

Then $S = N_m$.

Proof

We prove this result by contradiction. Assume that $S \neq N_m$. Then there exists a nonempty subset S' of N_m consisting of all the elements of N_m which do not belong to S. Since $S' \subseteq N_m$ and $S' \neq \varnothing$, it follows that the set S' has a least element, say k. By property (1), $m \in S$; therefore, $m \notin S'$, implying $k \neq m$ and $k > m$. Because k is the least element of S' and since $k - 1 \geq m$, it follows that $k - 1 \in N_m$ and $k - 1 \notin S'$, so $k - 1 \in S$. Using the fact that $k - 1 \in S$, we conclude, by property (2), that $(k - 1) + 1 = k \in S$. However, then $k \in S'$ and $k \in S$, which is impossible. Thus, we have arrived at a contradiction. This implies that our initial assumption that $S \neq N_m$ is incorrect, yielding the desired result $S = N_m$. ∎

With the aid of Theorem A.6, we can now establish the result we are primarily interested in.

Theorem A.7
First Principle of Mathematical Induction

Let m be an integer, and let $S_m, S_{m+1}, S_{m+2}, \ldots$ be statements with the properties:

(1) *S_m is a true statement;*

(2) *If S_k is a true statement, then S_{k+1} is a true statement.*

Then each of the statements S_m, S_{m+1} $S_{m+2}, \ldots$ is true.

Proof

We define a set S of integers as follows:

$$S = \{i \mid i \geq m, S_i \text{ is a true statement}\}.$$

Since S_m is a true statement, it follows that $m \in S$. If $k \in S$, then, necessarily, S_k is true. However, by property (2), S_{k+1} is also true, which implies that $k + 1 \in S$; that is, if $k \in S$, then $k + 1 \in S$. Therefore, because $S \subseteq N_m$ and S satisfies the two properties of Theorem A.6, we conclude, by Theorem A.6, that $S = N_m$. Hence, for every $i \geq m$, S_i is a true statement. ∎

The First Principle of Mathematical Induction is often used for the case $m = 1$. We state this special case:

Corollary
First Principle of Mathematical Induction

Let $S_1, S_2, S_3, \ldots$ be statements with the properties:

(1) *S_1 is a true statement;*

(2) *If S_k is a true statement, then S_{k+1} is a true statement.*

Then each of the statements $S_1, S_2, S_3, \ldots$ is true.

As a simple (and common) illustration of the First Principle of Mathematical Induction, we consider the mathematical statement

$$S_n:\quad 1 + 3 + 5 + \cdots + (2n - 1) = n^2.$$

We show that S_n is a true statement for every positive integer n. First, since $1 = 1^2$, S_1 is true. Next, suppose that

$$S_k:\quad 1 + 3 + 5 + \cdots + (2k - 1) = k^2$$

is a true statement. We show that

$$S_{k+1}:\quad 1 + 3 + 5 + \cdots + (2k - 1) + (2k + 1) = (k + 1)^2$$

is a true statement. Since $1 + 3 + 5 + \cdots + (2k - 1) = k^2$, it follows that

$$[1 + 3 + 5 + \cdots + (2k - 1)] + (2k + 1) = k^2 + (2k + 1);$$

however, $k^2 + (2k + 1) = (k + 1)^2$, so S_{k+1} is a true statement. Hence S_n is a true statement for every positive integer n.

A variation of the First Principle of Mathematical Induction is also very useful. Since its proof is similar to the proof of Theorem A.7, we state this result without proof.

Theorem A.8
Second Principle of Mathematical Induction

Let m be an integer, and let $S_m, S_{m+1}, S_{m+2}, \ldots$ be statements with the properties:

(1) *S_m is a true statement;*

(2) *If S_i is a true statement for each i, $m \le i \le k$, then S_{k+1} is a true statement.*

Then each of the statements $S_m, S_{m+1}, S_{m+2}, \ldots$ is true.

Problem Set A.6

33. Which of the following sets are well-ordered? (Q denotes the set of rational numbers, i.e., fractions.)

    (a) N

    (b) Z

    (c) $S = \{n | n \in Z, n \geq -10\}$

    (d) $S = \{n | n \in Q, n \geq -10\}$

    (e) the open interval $(-2, 2)$

    (f) the closed interval $[-2, 2]$

    (g) $S = \{-2, -1, 0, 1, 2\}$

    (h) $S = \{p | p \text{ is a prime}\} = \{2, 3, 5, 7, 11, 13, 17, \ldots\}$

    (i) $S = \{1/2, 1/3, 1/4, 1/5, \ldots\}$

    (j) $S = \{1/2, 2/3, 3/4, 4/5, \ldots\}$

    (k) $S = \{x | x \in Q, -1 \leq x \leq 1\}$.

34. Prove that if A is any well-ordered set of numbers and B is a nonempty subset of A, then B is also well-ordered.

35. Prove that the statement

$$S_n\colon\quad 1 + 2 + 3 + \cdots + n = n(n + 1)/2$$

is true for every positive integer n.

36. Prove that

$$S_n\colon\quad 1^2 + 2^2 + 3^2 + \cdots + n^2 = n(n + 1)(2n + 1)/6$$

is true for every positive integer n.

37. Prove that

$$1^3 + 2^3 + 3^3 + \cdots + n^3 = n^2(n + 1)^2/4$$

for every positive integer n.

38. Prove that

$$1^2 + 3^2 + 5^2 + \cdots + (2n - 1)^2 = (4n^3 - n)/3$$

for every positive integer n.

39. Prove that

$$\frac{1}{1 \cdot 2} + \frac{1}{2 \cdot 3} + \cdots + \frac{1}{n(n+1)} = \frac{n}{n+1}$$

for every positive integer n.

40. Prove that

$$1 \cdot 3 + 2 \cdot 4 + 3 \cdot 5 + \cdots + n(n + 2) = n(n + 1)(2n + 7)/6$$

for every positive integer n.

41. Prove using mathematical induction that the total number of distinct subsets of a set with n elements is 2^n.

42. State Theorem A.8 for the special case $m = 1$.

43. Prove Theorem A.8.

44. Define $x_1 = 1$, $x_2 = 2$, and $x_{n+2} = x_{n+1} + x_n$ for $n \geq 1$.
    (a) Prove that x_n is defined for every positive integer n.
    (b) Prove that $4^n x_n < 7^n$ for every positive integer n.

45. Prove that $2^n > n$ for every integer $n \geq 0$.

46. Prove the following implication for every positive integer n: If $x_1, x_2, \ldots, x_n$ are any n numbers such that $x_1 \cdot x_2 \cdots x_n = 0$, then at least one of the numbers $x_1, x_2, \ldots, x_n$ is 0. (Use the result that if the product of two numbers is 0, then at least one of the numbers is 0.)

47. Prove that $n! > 2^n$ for every integer $n \geq 4$. [*Note*: $n! = n(n - 1)(n - 2) \cdots 3 \cdot 2 \cdot 1$.]

48. Prove that $2^n > n^3$ for every integer $n \geq 10$.

49. It is known that if an integer $n\,(\geq 2)$ is not prime, then there exist integers a and b with $1 < a < n$ and $1 < b < n$ such that $n = ab$. Prove that every integer $n \geq 2$ is either prime or a product of primes.

50. If $n \geq 2$, prove that the number of prime factors of n is less than $2 \ln n = 2 \log_e n$.

51. Prove that for $n \geq 2$, the maximum number of points of intersection of n distinct lines in the plane is $n(n-1)/2$.

52. Let $f(x) = \ln x = \log_e x$. Prove that for every positive integer n, the nth derivative of $f(x)$ is given by

$$f^{(n)}(x) = \frac{(-1)^{n+1}(n-1)!}{x^n}$$

53. Let $f(x) = xe^{-x}$. Prove that $f^{(n)}(x) = (-1)^n e^{-x}(x-n)$ for every positive integer n.

54. Let U be a universal set. Prove that if $A_1, A_2, \ldots, A_n$ $(n \geq 2)$ are any subsets of U, then

$$\overline{A_1 \cup A_2 \cup \cdots \cup A_n} = \bar{A}_1 \cap \bar{A}_2 \cap \cdots \cap \bar{A}_n.$$

55. Let U be a universal set. Prove that if $A_1, A_2, \ldots, A_n$ $(n \geq 2)$ are any subsets of U, then

$$\overline{A_1 \cap A_2 \cap \cdots \cap A_n} = \bar{A}_1 \cup \bar{A}_2 \cup \cdots \cup \bar{A}_n.$$

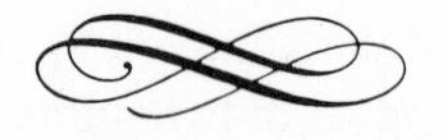

Answers, Hints, and Solutions to Selected Exercises

Chapter 1

Problem Set 1.2
(page 8)

5. The formula probably does not hold for any other values of n.

7. $dT/dt = k(T - 70)$.

11. What value of x produces the maximum profit?

13. You could list the results of the various tosses by $R_1R_2R_3R_4$, where $R_i (i = 1, 2, 3, 4)$ is the result on toss i. Hence, $R_i = H$ (heads) or T (tails) for $i = 1, 2, 3, 4$. The total number of possible "combinations" is 16. Of these, 6 give two heads and two tails, and 10 do not. So, on the average, 6/16 of the time you win \$11, while 10/16 of the time you pay \$10. The main question to ask is whether you are likely to win or lose money.

Problem Set 1.3
(page 12)

15.

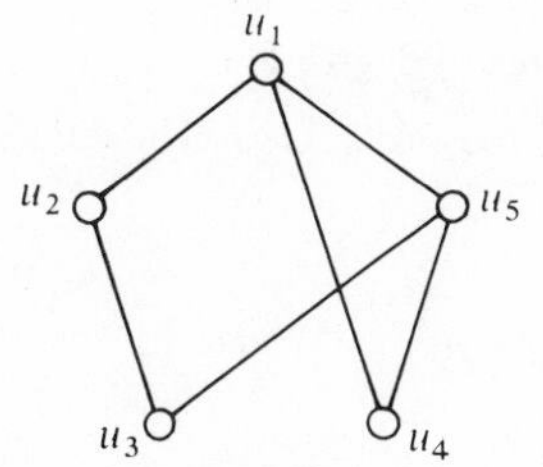

$$R = \{(u_1, u_2), (u_2, u_1), (u_1, u_4), (u_4, u_1), (u_1, u_5), (u_5, u_1), (u_2, u_3), (u_3, u_2), (u_3, u_5), (u_5, u_3), (u_4, u_5), (u_5, u_4)\}.$$

17. 0, 1, 2, 3.

19. Yes; let $V = \{u_1, u_2, u_3\}$ and $E = \{u_1u_2, u_1u_3, u_2u_3\}$.
No; for if $V = \{u_1, u_2, u_3, u_4\}$, then $u_1u_2, u_3u_4 \in E$, but u_1u_2 and u_3u_4 are not adjacent.

21. $n - 1$.

23. See the example in Exercise 19.

Problem Set 1.4
(page 15)

27. For any two vertices, determine whether it is possible to proceed from one to the other by a sequence of edges.

29. For graph G_1, let $V(G_1)$ represent the Student Council members. Join two vertices of G_1 by an edge if and only if the two individuals belong to the same committee. For graph G_2, let $V(G_2)$ represent the ten committees. Join two vertices of G_2 by an edge if and only if the two committees have some members in common. (There is at least one other natural graph G_3 possible here.)

Problem Set 1.5
(page 18)

33. (i) 6; (ii) 12; (iii) 20; (iv) $n(n - 1)$.

35. You should be able to get from any place to any other place. Yes, this could be determined from the digraph.

Chapter 2

Problem Set 2.1
(page 30)

1. $\deg v_6 = 1; \deg v_1 = \deg v_7 = \deg v_8 = 2; \deg v_2 = \deg v_3 = \deg v_4 = 3;$
$\deg v_5 = 4.$ $p = 8, q = 10.$

3. A graph of order 5 cannot have a vertex of degree 5.

5. Yes; the number of degrees listed is the order of G. The sum of the degrees is twice the size of G.

7. (a) K_3 (b) K_2 (c) none (d) K_1 (e) K_2
(f) the graph having order two and size zero.

9. $q = pr/2$; for K_p, $q = p(p - 1)/2$.

11. K_{m+1} or K_{n+1}. If m and n are both required degrees, let G be the "disjoint" union of K_{m+1} and K_{n+1}.

13. *Hint*: Represent this situation by a graph where the vertices correspond to the people and each edge represents a handshake. Now consider the degrees of this graph.

Problem Set 2.2
(page 38)

15. (a) $\phi(u_1) = u_2$, $\phi(v_1) = v_2$, etc.
    (b) $\theta(u_2) = u_1$, $\theta(v_2) = v_1$, etc.

17. No; otherwise, this would contradict Theorem 2.4.

19. There are several ways of doing this. One possible hint: F_1 contains pairs of nonadjacent vertices (such as u_1 and u_4) which are mutually adjacent to two vertices. F_2 does not have this property.

21. *Hint*: G_3 is isomorphic to G_1.

23. *Additional Hint*: Assume that each subset S_i of S has less than $\{n/k\}$ elements. Now show that S has less than n elements.

25. *Hint*: If a graph has order 4 and size 2, then the two edges may be adjacent or nonadjacent.

Problem Set 2.3
(page 43)

27. 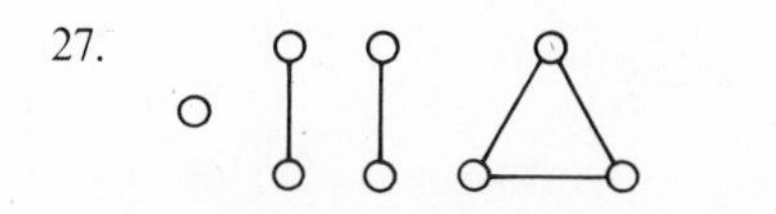

29. 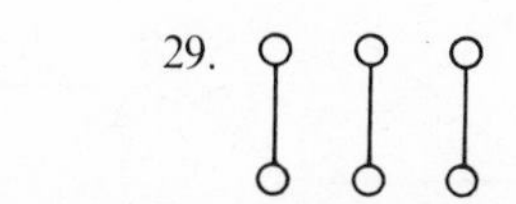

31. No; every component must have at least one vertex.

33. *Additional Hint*: If K_x, $1 \leq x \leq p - 1$, is one component, then K_{p-x} is the other component. The size of G is $f(x) = x(x-1)/2 + (p-x)(p-x-1)/2$. Now, minimize $f(x)$.

35. *Hint*: If G is connected, then for every two vertices u and v there is a u-v path, and therefore a u-v walk. To prove the converse statement, show that if there exists a u-v walk in G, then G must contain a u-v path.

37. *Additional Hint*: Let u_1 be a vertex in a component G_1 of G. Since $\deg u_1 \geq (p-1)/2$, G_1 has order at least $1 + (p-1)/2 = (p+1)/2$.

39. One possible circuit is $C: v_5, v_4, v_2, v_5, v_6, v_3, v_5$.

$$V(H) = \{v_2, v_3, v_4, v_5, v_6\};$$

$$E(H) = \{v_2v_4, v_2v_5, v_4v_5, v_3v_5, v_3v_6, v_5v_6\}.$$

41. *Hint*: See Exercise 35.

43. Neither; no edge is repeated in a circuit or cycle.

Problem Set 2.4
(page 47)

45. Cut-vertices: v_4, v_5, v_6. Bridges: v_1v_4, v_5v_6.

47. (a) By the Pigeonhole Principle, there is a component containing $\{11/2\} = 6$ vertices.
    (b) Let $V(G) = \{v, v_1, v_2, \ldots, v_{10}\}$ and $E(G) = \{vv_1, vv_2, \ldots, vv_{10}\}$.

49.

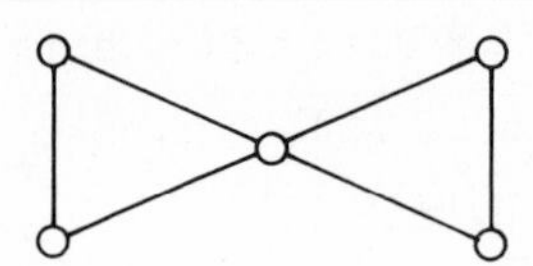

51. Yes. (*Hint*: First show that if e lies on a circuit of G, then e lies on a cycle of G.)

53.

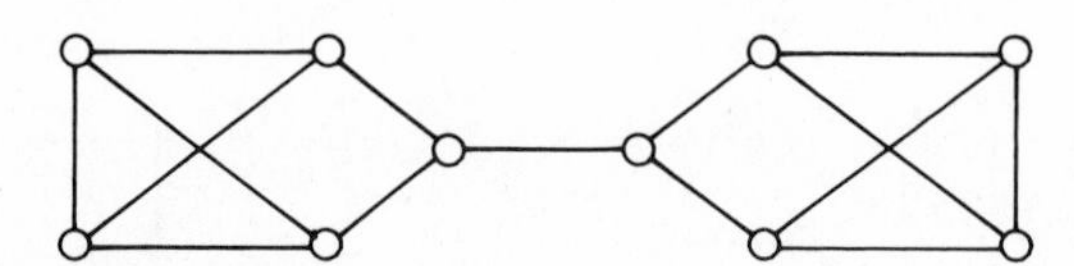

55. See the solution of Exercise 49.

57.

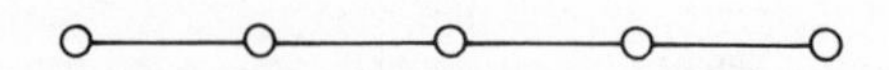

59. *Hint*: Let $e = uv$. Since G is not isomorphic to K_2, at least one of u and v has degree at least two. Suppose $\deg u \geq 2$. The graph $G - e$ has two components, one containing u and the other containing v. Show that $G - u$ is disconnected.

Chapter 3

Problem Set 3.1
(*page 61*)

1. G_1 traversable; G_2 neither; G_3 traversable; G_4 neither; G_5 eulerian.
3. G is traversable.
5. A connected multigraph G with exactly four odd vertices contains two edge-disjoint trails such that every edge of G belongs to one of these trails. (*Hint for proof*: Let u and v be two distinct odd vertices of G, and consider $G + uv$.)
7. Yes. (*Hint*: Model this situation with a graph.)
9. No. (*Hint*: Model this situation with a graph.)
11. The murderer is (gasp)—James Bomb! The graph which models this problem is traversable; however, the vertex corresponding to the exterior of the house is even. This implies that it is not possible to enter the house, go through each door exactly once, and then leave the house. Hence, the gardener lied. However, this does not prove that the gardener killed the Count. Indeed, from the information available, the only way James Bomb could possibly know who murdered the Count is if he committed this terrible crime himself.
13. Answer to first question: No.
15. *Hint*: If G is connected and its edge set can be partitioned into cycles, then each vertex v of G belongs to a certain number of cycles. What can be said about deg v? To show the converse statement, let G be an eulerian graph. Thus, G contains an eulerian circuit. Now use the fact that every circuit contains a cycle.

Problem Set 3.2
(*page 73*)

17. (a)

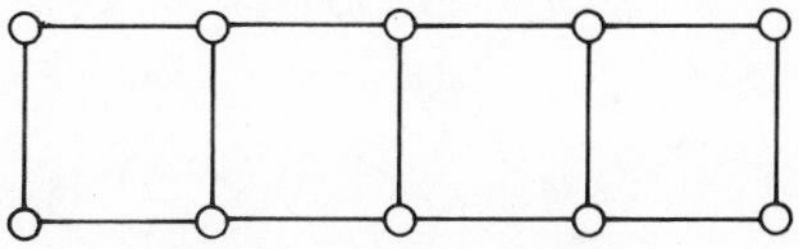

(b)

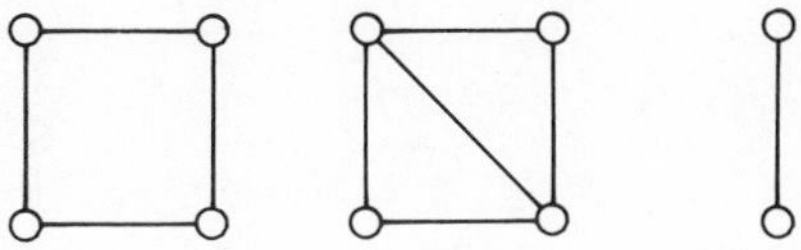

19. *Hint*: Find a hamiltonian cycle in the graph of the cube.

21. Consider the following graph G. Let $H = K_{(p-1)/2}$. Define G to be two disjoint copies of H together with a new vertex v adjacent to all vertices in both copies of H.

23. False. To verify this, an example must be given.

25. *Additional Hint*: The hamiltonian cycle of G cannot contain two consecutive vertices labeled M or two consecutive vertices labeled W.

27. The graph of Figure 3.21 is not hamiltonian. (*Hint*: Assume this graph is hamiltonian, and obtain a contradiction.)

29. *Hint*: Consider two cases, depending on whether mn is even or odd.

31. Beginning with city A, the hamiltonian cycle obtained by always proceeding to the next city using the least expensive route is A, B, C, D, E, A. The cost of this cycle is \$270. Now determine four other cycles, one beginning at each of B, C, D, and E, respectively. Show that there exists a hamiltonian cycle whose cost is less than the cost of each of these five cycles.

Chapter 4

Problem Set 4.1
(page 87)

1. (a) 1 (b) 1 (c) 2 (d) 3 (e) 6.
The six trees of order 6 are

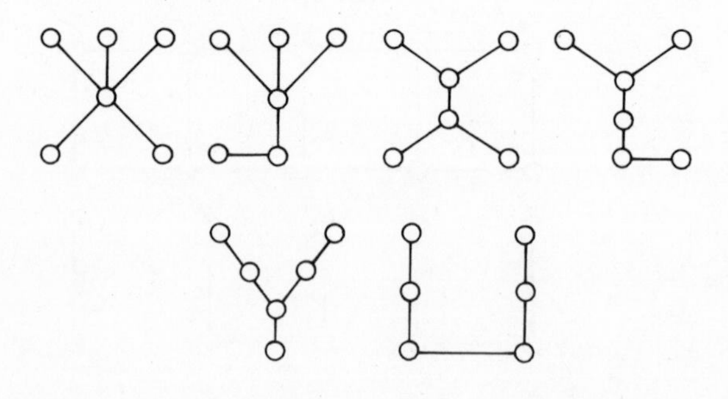

3. Yes. K_1 and K_2 are the only regular trees.

5.

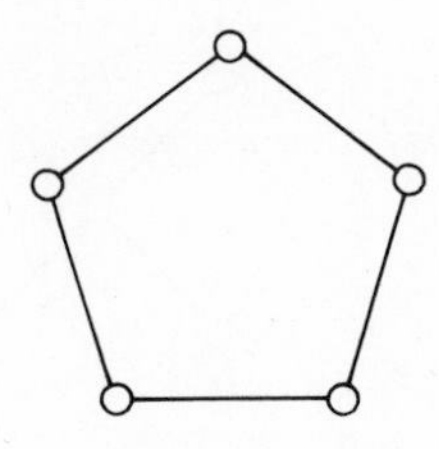

7. Let $G_1, G_2, \ldots, G_k$ be the k components of G. Thus, each component is a tree. If G_i has order p_i and size q_i, for each $i = 1, 2, \ldots, k$, then $p_1 + p_2 + \cdots + p_k = p$, $q_1 + q_2 + \cdots + q_k = q$, and $q_i = p_i - 1$. Therefore,

$$\begin{aligned} q &= q_1 + q_2 + \cdots + q_k \\ &= (p_1 - 1) + (p_2 - 1) + \cdots + (p_k - 1) \\ &= p - k. \end{aligned}$$

9. If G has no cycles, then G is a tree. Next, suppose that G has cycles. We can obtain a tree T by deleting edges from G; namely, delete an edge e_1 on a cycle of G, producing $G - e_1$. If $G - e_1$ is not a tree, then we select an edge e_2 on a cycle of G, producing $G - e_1 - e_2$. We continue this procedure until a tree T is produced. But T has order p and less than $p - 1$ edges. This is impossible for a tree.

11. Since G is a forest, G has no cycles. Assume that G is not a tree. Then G has components $G_1, G_2, \ldots, G_k$, where $k \geq 2$. By Exercise 7, $q = p - k$. Since $q = p - 1$, it follows that $k = 1$, giving us a contradiction.

13. (a) The four economy trees are:

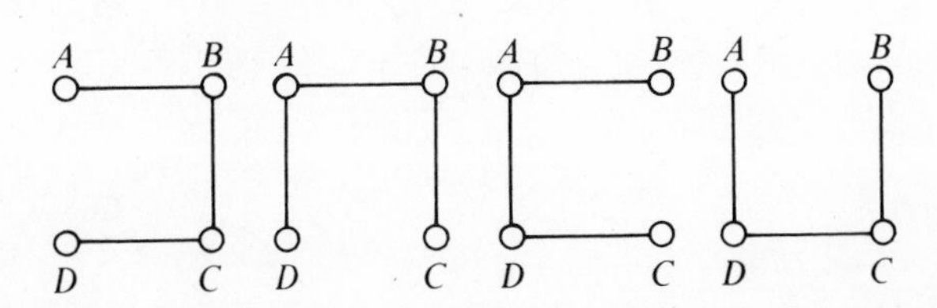

One possible decision is not to have track between that pair of cities among $\{A, B\}$, $\{B, C\}$, $\{C, D\}$, and $\{D, A\}$ which ordinarily has the least railroad traffic.

(b) (i) Construct track between E and any one of A, B, C, and D.
(ii)

(c) Yes.

17. *Hint*: Represent this situation by a graph whose vertices correspond to the prison cells and the exterior. An edge represents the existence of a gate between the appropriate cells, or a cell and the exterior.

Problem Set 4.2
(page 95)

19. (a) 1/6 (b) 1/6

21. Answer for (b): 6/16.

23.

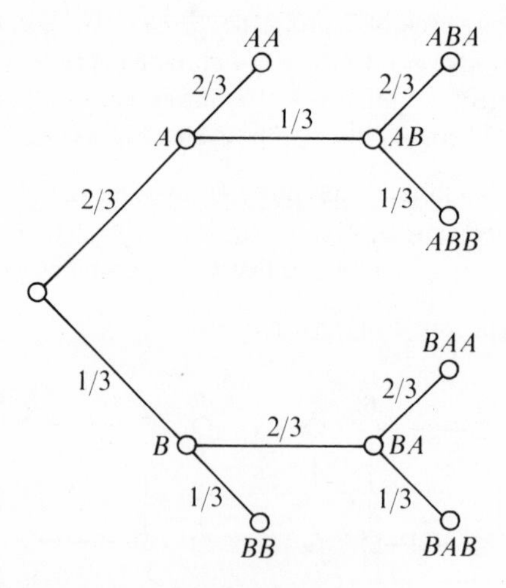

The probability of A winning the playoff is

$$\frac{2}{3}\cdot\frac{2}{3}+\frac{2}{3}\cdot\frac{1}{3}\cdot\frac{2}{3}+\frac{1}{3}\cdot\frac{2}{3}\cdot\frac{2}{3}=\frac{20}{27}.$$

Problem Set 4.3
(*page 102*)

25. The critical path depends on the times assigned.
27. Label the digraph as follows:

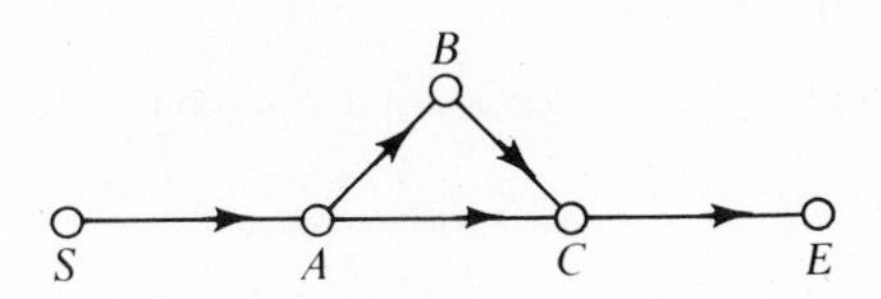

If this is an activity digraph, then C may start immediately after A is completed. However, C cannot start until B is completed, and B cannot start until A is completed. Thus, C cannot start after A is completed.

Chapter 5

Problem Set 5.1
(*page 113*)

1.

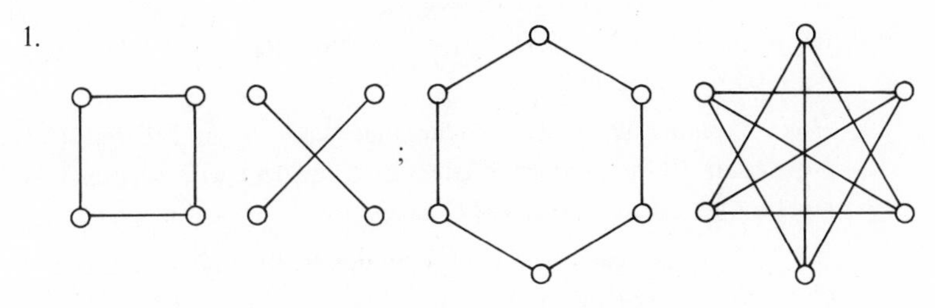

3. $9 - d_1, 9 - d_2, \ldots, 9 - d_{10}$.
5. A cycle of order 5.
7. (a) A path of order 4.
   (b) K_2 has one edge; K_3 has three edges. Since these complete graphs have odd size, each graph G of order 2 or 3 has size unequal to that of $\overline{G}$, so G is not isomorphic to $\overline{G}$.
9. It means that there exists a positive integer p such that for any graph G of order p, either G contains K_m or $\overline{G}$ contains K_n.
11. K_6 is a subgraph of K_7. Since $r(K_3, K_3) = 6$, K_6 contains a red K_3 or a blue K_3. Therefore, K_7 contains a red K_3 or a blue K_3.

13. (a) $r(P_3, P_3) = 3$. *Proof*: $r(P_3, P_3) > 2$ since no matter how the edge of K_2 is colored, K_2 contains neither a red P_3 nor a blue P_3. If the edges of K_3 are colored red or blue, at least two of the three edges are colored the same. If red, then a red P_3 is produced; otherwise, a blue P_3 is produced.

(b) $r(P_3, K_3) = 5$.

15. Six people. (*Hint*: Represent the situation as a graph.)

Problem Set 5.2
(*page 120*)

17.

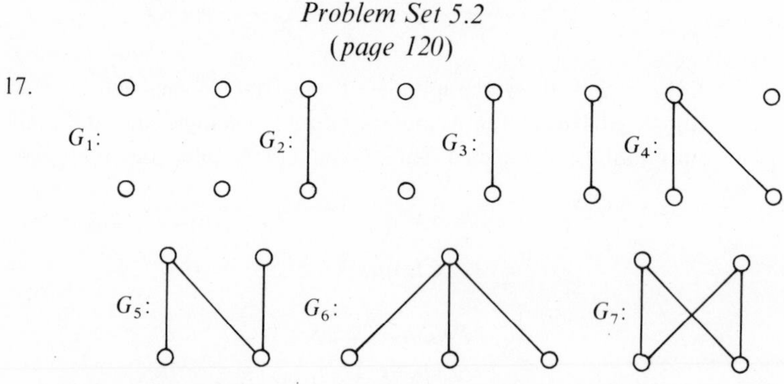

19. *Hint*: If the cardinalities of two vertex subsets are x and $p - x$, then the maximum size is $f(x) = x(p - x)$. Now maximize $f(x)$ over all permissible values of x.

21. *Hint*: Assume that G is r-regular; that $|V_1| = m$; and that $|V_2| = n$. Now count (i) the number of edges of G incident with vertices in V_1, and (ii) the number of edges of G incident with vertices in V_2.

23. *Hint*: Consider each vertex of V_1 individually, and investigate the possibility of a function $f: V_1 \rightarrow V_2$.

25. *Hint*: Represent the situation as a graph and apply Theorem 5.4.

27. Five applicants.

Chapter 6

Problem Set 6.1
(*page 131*)

1. *One possible argument*: The pseudograph of Figure 6.5 has 12 edges. If this pseudograph contained three edge-disjoint, 2-regular spanning sub-pseudographs, then they would necessarily use all 12 edges.

One of these sub-pseudographs contains the loop (labeled 1) at W and another contains the loop (labeled 2) at G. There do not exist two edges joining R and B, one labeled 3 and one labeled 4, so that these two loops belong to distinct sub-pseudographs. Hence, the sub-pseudograph containing the loop at W must contain a triangle on R, B, and G whose three edges are labeled 2, 3, and 4. However, there is no edge joining R and G. Contradiction!

3. No solution. (*Hint*: Show that there is no 2-regular spanning sub-pseudograph of the pseudograph which contains the edge labeled 4 joining W and B. Then show that there is none containing the edge labeled 4 joining R and B.)

5. See Exercise 3.

7. No.

9. No.

11. 41,472.

Problem Set 6.2
(*page 135*)

13. Six moves.

15. There is no knight's tour of the 4-by-5 chessboard. To see this, begin with the four corner squares.

Problem Set 6.3
(*page 139*)

19. The hamiltonian cycle is (0, 0, 0), (1, 0, 0), (1, 1, 0), (0, 1, 0), (0, 1, 1), (1, 1, 1), (1, 0, 1), (0, 0, 1), (0, 0, 0).

Problem Set 6.4
(*page 144*)

21. *Hint*: We can show this by considering cases. Necessarily, any trail from (3, 3) to (0, 0) which alternates positive and negative edges must first enter (1, 0) or (2, 0) by means of a positive edge. Determine the shortest such trail from (3, 3) to (2, 0) and from (2, 0) to (0, 0).

23. *Hint*: One possibility is to get all the women into town first.

Answers, Hints, and Solutions to Selected Exercises

Chapter 7

Problem Set 7.1
(page 152)

1. If G is hamiltonian, we may select the cycle C in the proof as a hamiltonian cycle. In this case, every vertex of G lies on C, and it is not necessary to consider the situation where there are vertices not belonging to C.

3. Since the graph G of Figure 7.3 contains no bridges, it follows, by Theorem 7.1, that G is orientable.

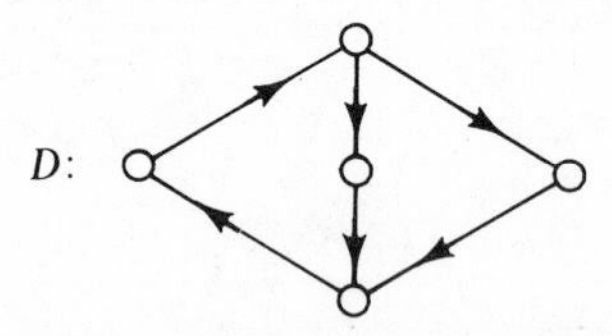

5. Yes. A graph with cut-vertices need not contain bridges.

7. The street system can be modeled by the following graph G. Since G contains no bridges, G is orientable. The strongly connected digraph D is obtained by assigning directions to the edges of G.

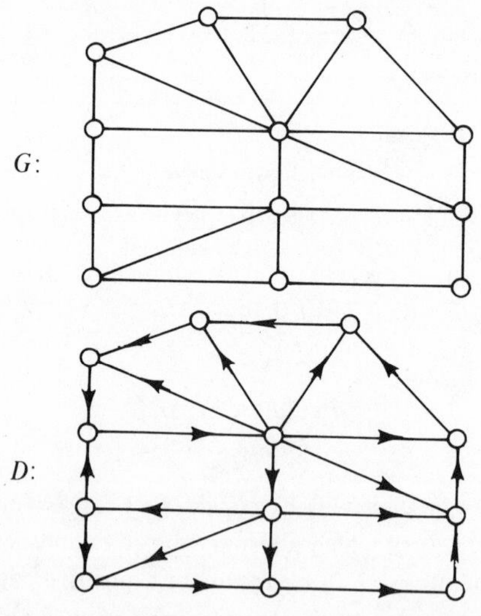

9. Let u be a vertex of G. Direct the edges of G incident with u toward u, and direct any other edges of G in any manner. The resulting digraph D

is *not* strongly connected, since there is no path from u to any other vertex of D.

Problem Set 7.2
(page 159)

11. A tournament T_1 is isomorphic to a tournament T_2 if there exists a one-to-one mapping ϕ from the vertex set of T_1 onto the vertex set of T_2 such that for every two distinct vertices u and v of T_1, (u, v) is an arc of T_1 if and only if $(\phi(u), \phi(v))$ is an arc of T_2.

13. (a), (b) $\sum_{i=1}^{p} \text{od } v_i = \sum_{i=1}^{p} \text{id } v_i = q$.
    (c) $p(p-1)/2$.

15. Exactly one of the arcs (u, v) and (v, u) belongs to T. If (u, v), for example, is an arc of T, then $d(u, v) = 1$ and $d(v, u) \geq 2$.

17. (a) It is possible for each team to win twice and lose twice, as shown in the following tournament:

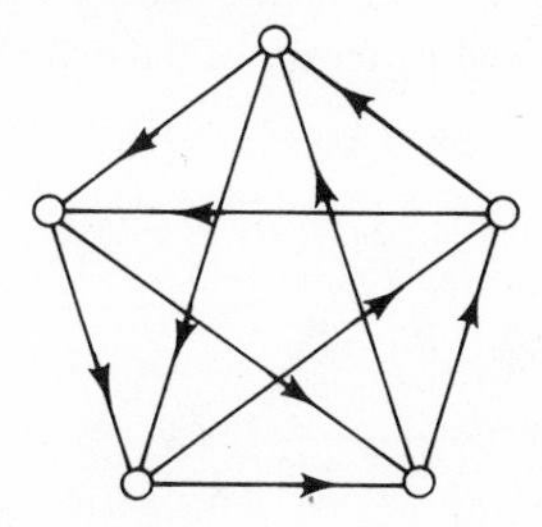

    (b) Suppose it is possible for all six teams to tie for first place. Then each team has the same number n of victories. In the entire tournament, the total number of victories is $6n = 15$, implying that $n = 2\frac{1}{2}$, which is impossible.

19. Examples with a five-team tournament are given below.

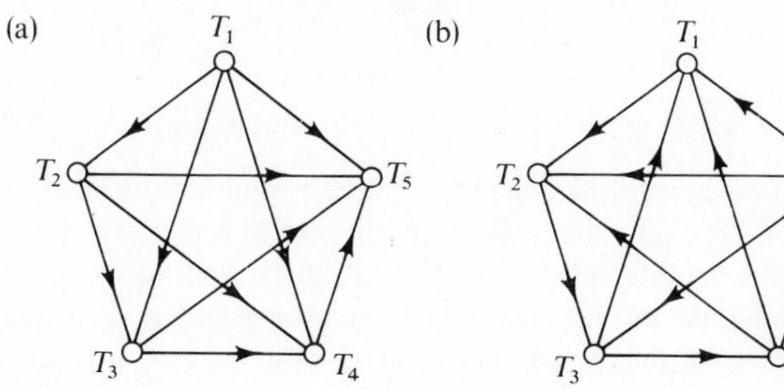

Problem Set 7.3
(*page 167*)

25. Councilman Adams would have been elected mayor, since he would have received more votes than the other two candidates. We could describe this by a tournament where (u, v) is an arc if candidate u receives more votes than candidate v.

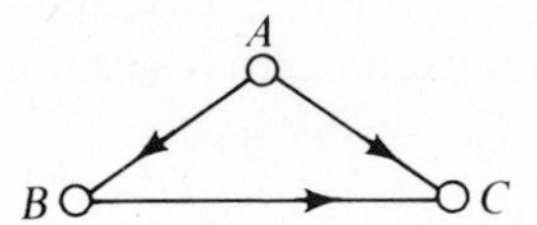

27. Councilman Carter would have been elected mayor. Since Adams was least preferred by the voters, he would have been eliminated. Between Beane and Carter, Councilman Carter would receive more votes. This could be described by means of the two tournaments:

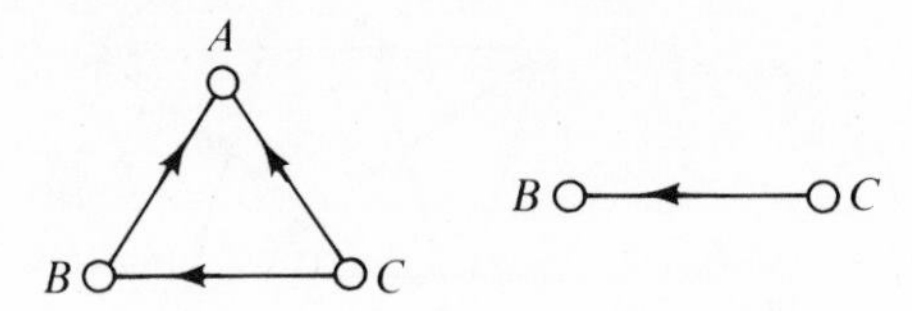

29. Councilman Carter would have been elected mayor.

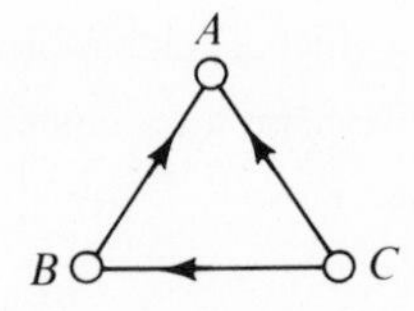

31. If nine broadcasters rate State the best team, then their votes would give 90 points to State. Similarly, they would give 81 points to Central. If the tenth broadcaster ranks Central first and does not rank State among the top ten teams, then Central would receive 91 points and State 90 points, giving Central first place in the final rankings.

Chapter 8

Problem Set 8.1
(*page 177*)

1. If the social system contains only one or two people, then the system is balanced. If the system contains more than two people, it is unbalanced.

3. S is balanced. If S contains no cycles, then S satisfies the condition (trivially) that each of its cycles is positive. Therefore, by Theorem 8.2, S is balanced.

5. Since each of the signed graphs $S_1, S_2, \ldots, S_n$ is balanced for each i ($1 \leq i \leq n$), the vertex set $V(S_i)$ can be partitioned as $V_{i1} \cup V_{i2}$, say, where each edge joining two vertices of V_{i1} or two vertices of V_{i2} is positive and each edge joining a vertex of V_{i1} and a vertex of V_{i2} is negative. Then, it is possible to partition $V(S)$ as $V_1 \cup V_2$, where

$$V_1 = V_{11} \cup V_{21} \cup \cdots \cup V_{n1}$$

and

$$V_2 = V_{12} \cup V_{22} \cup \cdots \cup V_{n2}.$$

Further, every edge joining two vertices of V_1 or two vertices of V_2 is positive, while each edge joining a vertex of V_1 and a vertex of V_2 is negative. Therefore, S is balanced.

7. Let P' be a positive u-v path and let P'' be a negative u-v path. Suppose P' is the path $w_0, w_1, \ldots, w_n$, where $w_0 = u$ and $w_n = v$. The path P'' also begins with vertex w_0 and terminates at w_n. Since P' and P'' are different paths, there must be a vertex w_i such that P' and P'' both begin with $w_0, w_1, \ldots, w_i$ but the vertex of P'' following w_i is not w_{i+1}. Clearly, the w_0-w_i subpaths of P' and P'' contain the same number of negative edges. Let the next vertex of P' which also belongs to P'' be w_j, where $j > i$. The w_i-w_j subpath of P' and the w_i-w_j subpath of P'' produce a cycle C_1 in S. If C_1 contains an odd number of negative edges, then C_1 is a negative cycle. Otherwise, the w_i-w_j subpath of P' and the w_i-w_j subpath of P'' both contain an odd number or both contain an even number of negative edges. Therefore, the w_0-w_j subpath of P' and the w_0-w_j subpath of P'' both contain an odd number or both contain an even number of negative edges. So, exactly one of the w_j-w_n subpath of P' and the w_j-w_n subpath of P'' is even and the other is odd. We may

proceed as before to produce a cycle C_2 in S. If C_2 is not negative, there exists a vertex $w_k (k > j)$ on both P' and P'' so that exactly one of the w_k-w_m subpath of P' and the w_k-w_n subpath of P'' is even and the other is odd. Since paths have a finite number of vertices, eventually we must arrive at a cycle C_m which is negative.

Problem Set 8.2
(page 181)

11. Let $S_1, S_2, \ldots, S_n$ be the components of S, where each of the components is clusterable. For $i = 1, 2, \ldots, n$, denote the clusters of S_i by $V_{i1}, V_{i2}, \ldots, V_{in_i}$. Then

$$V_1 = V_{11} \cup V_{21} \cup \cdots \cup V_{n1},$$
$$V_2 = V_{12} \cup V_{22} \cup \cdots \cup V_{n2},$$

etc. are clusters of S. The minimum number of clusters possible for S is the maximum of the minimum number of clusters of its components.

Chapter 9

Problem Set 9.1
(page 200)

1.

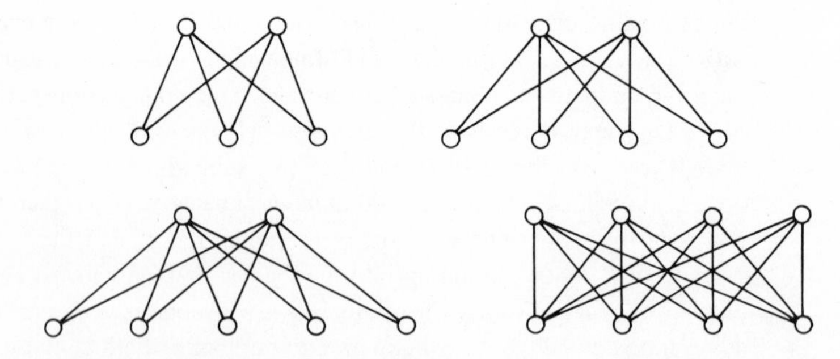

3. For the region R, the vertices of the boundary are v_1, v_2, v_4, v_5, v_6, and the edges of the boundary are $v_1v_2, v_2v_6, v_5v_6, v_1v_5$, and v_4v_5. For the

exterior region, the vertices of the boundary are all of $V(G)$ except v_4 and the edges of the boundary are all of $E(G)$ except $v_4 v_5$.

5. Suppose G has order p and size q. If G is drawn as a plane graph in any manner whatsoever, then by Theorem 9.1, $p - q + r = 2$, or $r = 2 - p + q$; that is, G has $2 - p + q$ regions.

7. (a) K_3 (b) $K(1, 2)$

9. False. The following graph G has $q = 3p - 6$, but G is nonplanar (since K_5 is a subgraph of G).

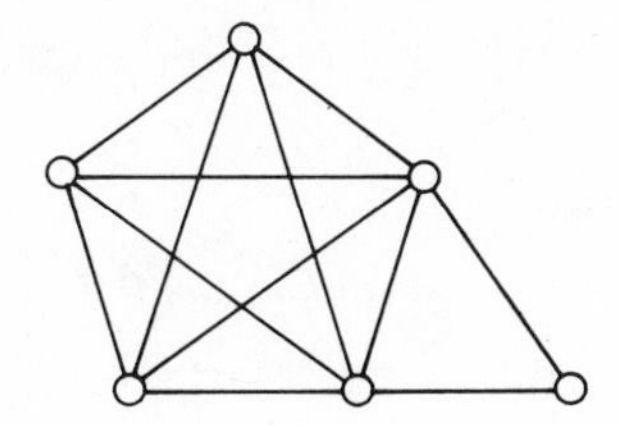

11. K_1, K_2, K_3, and K_4.

13. The Petersen graph is not planar. It contains a subgraph which is a subdivision of $K(3, 3)$.

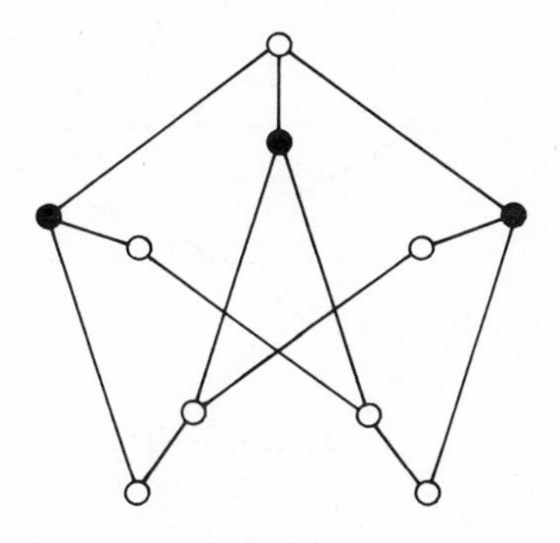

Problem Set 9.2
(*page 208*)

15. The chromatic number of a cycle is 2 or 3, depending on whether its order is even or odd.

17. 4, 5, $n + 1$; 2, 2, 2,

19. Label G_1 as indicated. Since v_1, v_2, and v_4 produce a triangle, three colors are required to color these vertices. Color v_1 with color 1, v_2 with color 2, and v_4 with color 3. If we color v_3 with 2, v_5 with 1, and v_6 with 3, we will have colored G_1 with three colors. Therefore, $\chi(G_1) = 3$.

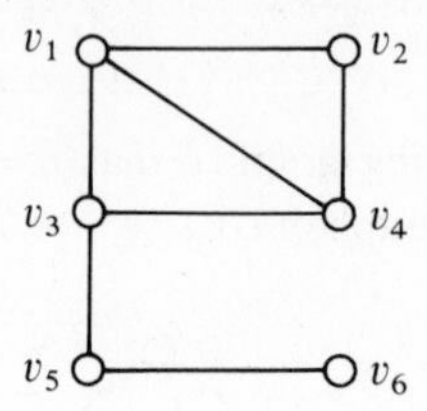

21. Four time periods.

Problem Set 9.3
(page 215)

23. 4

25.

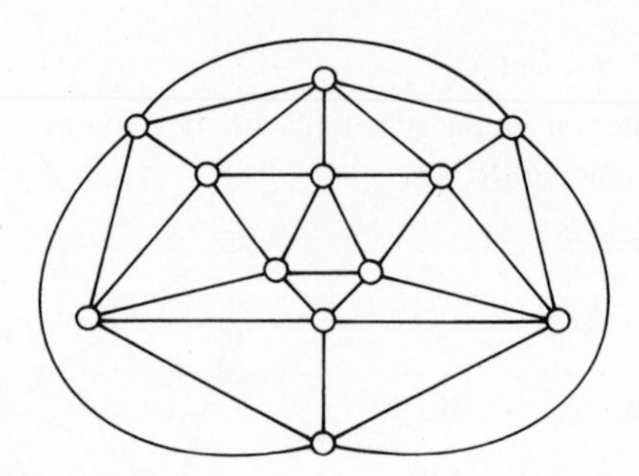

27. No; yes, $K(3, 3)$; yes

Chapter 10

Problem Set 10.1
(page 221)

1. v_1, v_2, v_3, v_4, v_5

$$A = \begin{bmatrix} 0 & 0 & 1 & 1 & 1 \\ 0 & 0 & 1 & 1 & 1 \\ 1 & 1 & 0 & 0 & 0 \\ 1 & 1 & 0 & 0 & 0 \\ 1 & 1 & 0 & 0 & 0 \end{bmatrix} \quad A^2 = \begin{bmatrix} 3 & 3 & 0 & 0 & 0 \\ 3 & 3 & 0 & 0 & 0 \\ 0 & 0 & 2 & 2 & 2 \\ 0 & 0 & 2 & 2 & 2 \\ 0 & 0 & 2 & 2 & 2 \end{bmatrix}$$

The (i, i) entry of A^2 gives the degree of v_i. The (i, j) entry, $i \neq j$, of A^2 gives the number of different v_i-v_j walks of length two.

3. Define a graph G by letting $V(G) = \{v_1, v_2, \ldots, v_p\}$. Then $v_i v_j$ is an edge of G if and only if the (i, j) entry of A is 1.

5. $a_{ij}^{(2)}$, $i \neq j$, is the number of v_i-v_j paths of length two in G, while $a_{ij}^{(2)} = \deg v_i$.

7. If G has order p and size q, then since the incidence matrix is square, $p = q$. By Exercise 4.10, G contains a cycle.

9. For the graph G of Figure 10.4,

$$B \cdot B^t = \begin{bmatrix} 2 & 1 & 0 & 1 \\ 1 & 3 & 1 & 1 \\ 0 & 1 & 2 & 1 \\ 1 & 1 & 1 & 3 \end{bmatrix}.$$

Hint to second question: Consider $A(G)$ and the degrees of G.

Problem Set 10.2
(*page 226*)

11. 5

13. There are several types of diagrams possible for such an embedding. One such diagram is:

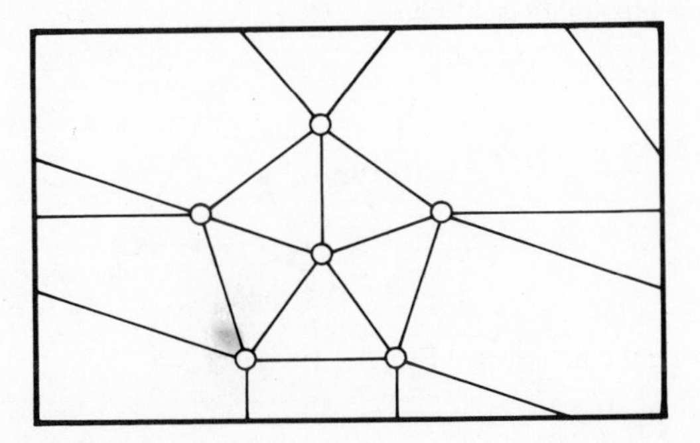

15. *Hint*: See Exercise 11.

17. *Hint*: If G is planar, no handles are necessary. Otherwise, draw G on the sphere, where, then, certain pairs of edges cross at points, say $P_1, P_2, \ldots, P_n$. Consider these n points.

Problem Set 10.3
(*page 235*)

19. Label the vertices of K_2 by 1, 2. Then $\mathscr{A}(K_2) = \{\alpha_0, \alpha_1\}$, where

$$\alpha_0 = \begin{pmatrix} 1 & 2 \\ 1 & 2 \end{pmatrix} \quad \text{and} \quad \alpha_1 = \begin{pmatrix} 1 & 2 \\ 2 & 1 \end{pmatrix}.$$

21. Label the vertex of degree 3 in $K(1, 3)$ by 1, and label the other three vertices by 2, 3, and 4. Then

$$\mathscr{A}(K(1, 3)) = \{\alpha_0, \alpha_1, \alpha_2, \alpha_3, \alpha_4, \alpha_5\},$$

where

$$\alpha_0 = \begin{pmatrix} 1 & 2 & 3 & 4 \\ 1 & 2 & 3 & 4 \end{pmatrix}, \quad \alpha_1 = \begin{pmatrix} 1 & 2 & 3 & 4 \\ 1 & 3 & 2 & 4 \end{pmatrix}, \quad \alpha_2 = \begin{pmatrix} 1 & 2 & 3 & 4 \\ 1 & 4 & 3 & 2 \end{pmatrix},$$

$$\alpha_3 = \begin{pmatrix} 1 & 2 & 3 & 4 \\ 1 & 2 & 4 & 3 \end{pmatrix}, \quad \alpha_4 = \begin{pmatrix} 1 & 2 & 3 & 4 \\ 1 & 3 & 4 & 2 \end{pmatrix}, \quad \alpha_5 = \begin{pmatrix} 1 & 2 & 3 & 4 \\ 1 & 4 & 2 & 3 \end{pmatrix}.$$

23. Let P_n: $1, 2, \ldots, n$. Then $\mathscr{A}(P_n) = \{\alpha_0, \alpha_1\}$, where

$$\alpha_0 = \begin{pmatrix} 1 & 2 & \ldots & n \\ 1 & 2 & \ldots & n \end{pmatrix} \quad \text{and} \quad \alpha_1 = \begin{pmatrix} 1 & 2 & \ldots & n-1 & n \\ n & n-1 & \ldots & 2 & 1 \end{pmatrix}.$$

25. *Hint*: First show that every automorphism of G is an automorphism of $\bar{G}$, and conversely.

27. One possibility is $\Delta = \{a\}$.

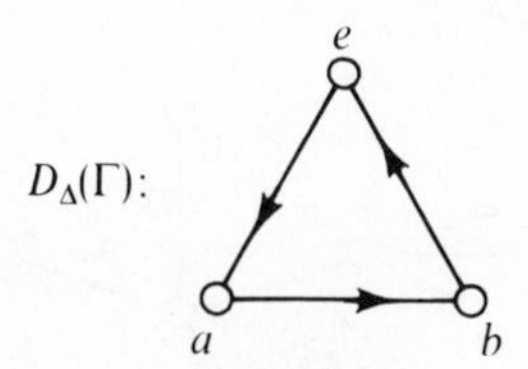

29. Using the notation of Exercise 19, we let $\Delta = \{\alpha_1\}$.

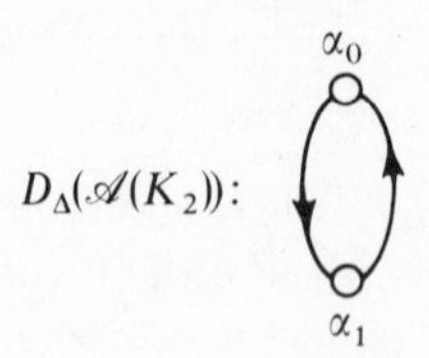

Appendix

Problem Set A.1
(*page 242*)

1. 8.

3. *Additional Hint*: If $x \in V_1 \cap W_1$, then $x \in V_1$ and $x \in W_1$. However, $V_1 \subseteq V$ and $W_1 \subseteq W$.

5. *Hint*: First show that $U \subseteq W$. Then show there exists $w \in W$ such that $w \notin U$.

Problem Set A.2
(*page 245*)

7. (a) All three properties (assuming that two coincident lines are parallel).
   (b) Reflexive and symmetric.
   (c) Transitive.

9. $R = \{(a, a), (a, b), (b, c)\}$ is one such example.

11. Symmetric and transitive.

Problem Set A.3
(*page 251*)

13. A person is born in the same year as himself (or herself). If person A is born in the same year as person B, then B is born in the same year as A. If A is born in the same year as B and B is born in the same year as C, then A is born in the same year as C. Each equivalence class consists of all people born in a particular year.

15. For each $a \in A$, $a\ R\ a$. (It was given that a and b need not be distinct.) Since every two elements are related, $a\ R\ b$ implies $b\ R\ a$, and $a\ R\ b$ and $b\ R\ c$ imply $a\ R\ c$. There is only one equivalence class, containing all elements of A.

17. The distinct elements of Z_2 are

$$[0] = \{0, \pm 2, \pm 4, \ldots\} \quad \text{and} \quad [1] = \{\pm 1, \pm 3, \pm 5, \ldots\}.$$

The distinct elements of Z_3 are

$$[0] = \{0, \pm 3, \pm 6, \ldots\}, \qquad [1] = \{1, -2, 4, -5, \ldots\},$$

and

$$[2] = \{2, -1, 5, -4, \ldots\}.$$

19. For each $a \in Z$, $a + a = 2a \equiv 0 \pmod 2$, since $2|2a$. Therefore, R is reflexive. If $a\,R\,b$, then $a + b \equiv 0 \pmod 2$; however, $a + b = b + a$, so $b + a \equiv 0 \pmod 2$, implying that R is symmetric. Assume $a\,R\,b$ and $b\,R\,c$. Then $a + b \equiv 0 \pmod 2$ and $b + c \equiv 0 \pmod 2$. Thus $2|(a + b)$ and $2|(b + c)$, so $a + b = 2m$ and $b + c = 2n$ for some integers m and n. Adding, we obtain $a + 2b + c = 2m + 2n$, or $a + c = 2(m + n - b)$. Hence $2|(a + c)$, so $a + c \equiv 0 \pmod 2$, implying that R is transitive. The distinct equivalence classes are $[0] = \{0, \pm 2, \pm 4, \ldots\}$ and $[1] = \{\pm 1, \pm 3, \pm 5, \ldots\}$.

21. *Hint*: For transitivity, assume $a^2 \equiv b^2 \pmod 5$ and $b^2 \equiv c^2 \pmod 5$, implying $a^2 - b^2 = 5m$ and $b^2 - c^2 = 5n$. Now add these two expressions. There are three distinct equivalence classes.

Problem Set A.4
(*page 255*)

23. *Hint*: Let $A_2 = B_2 = Z$. Define $f_2(n) = 2n$.

25. First we show that f^{-1} is a function. Let $y \in B$. We need to show that there is a unique $x \in A$ such that $(y, x) \in f^{-1}$. Since f is onto, there exists $x \in A$ such that $f(x) = y$. Since f is one-to-one, x is the only element of A whose image is y. Hence x is the only element of A such that $(y, x) \in f^{-1}$. To show f^{-1} is onto, let $a \in A$. Suppose $f(a) = b$. Then $f^{-1}(b) = a$, and f^{-1} is onto. To show f^{-1} is one-to-one, assume $f^{-1}(b_1) = f^{-1}(b_2)$. Suppose $f^{-1}(b_1) = a_1$ and $f^{-1}(b_2) = a_2$; thus, $a_1 = a_2$. So $f(a_1) = f(a_2)$, but since $f(a_1) = b_1$ and $f(a_2) = b_2$, we get $b_1 = b_2$.

Problem Set A.5
(*page 260*)

27. *Hint*: See Example A.1.

29. If $x = 3$, then $x^2 = (3)^2 = 9$. Direct proof.

31. If we win the game, then we are champions.
We win the game only if we are champions.
To be champions, it is sufficient to win the game.
To win the game, it is necessary to be champions.

Problem Set A.6
(page 264)

33. (a), (c), (g), (h), (j).

35. S_1 is true since $1 = 1(1 + 1)/2$. Assume S_k is true, i.e., assume $1 + 2 + \cdots + k = k(k + 1)/2$. Consider

$$\begin{aligned} 1 + 2 + \cdots + (k + 1) &= (1 + 2 + \cdots + k) + (k + 1) \\ &= k(k + 1)/2 + (k + 1) \\ &= [k(k + 1) + 2(k + 1)]/2 \\ &= (k + 1)(k + 2)/2, \end{aligned}$$

implying that S_{k+1} is true. By the First Principle of Mathematical Induction, S_n is true for every positive integer n.

37. *Hint*:

$$(1^3 + 2^3 + \cdots + k^3) + (k + 1)^3 = [k^2(k + 1)^2/4] + (k + 1)^3.$$

39. *Hint*:

$$k/(k + 1) + 1/(k + 1)(k + 2) = (k + 1)/(k + 2).$$

41. We can begin with $n = 0$. There is one subset of a set with no elements, namely $\varnothing$; that is, the number of subsets of a set with 0 elements is $2^0 = 1$. Assume the result is true for any set with k elements, where $k \geq 0$. Let $A = \{a_1, a_2, \ldots, a_k, a_{k+1}\}$ be an arbitrary set with $k + 1$ elements. We show that the number of subsets of A is 2^{k+1}. Let $B = \{a_1, a_2, \ldots, a_k\}$. Since B has k elements, there are 2^k subsets of B. Since $B \subseteq A$, each subset of B is a subset of A; in fact, every subset of A not containing a_{k+1} is a subset of B. Thus there are 2^k subsets of A not containing a_{k+1}. The subsets of A containing a_{k+1} are obtained by taking the union of the subsets of B with $\{a_{k+1}\}$; hence there are 2^k subsets of A containing a_{k+1}. Therefore, the total number of subsets of A equals $2^k + 2^k = 2^{k+1}$.

43. *Hint*: First prove the following result. Let S be a subset of N_m with the properties:
    (1) $m \in S$;
    (2) If $m, m + 1, \ldots, k$ are in S (where $k \geq m$), then $k + 1 \in S$.
    Then $S = N_m$.

45. The result is true for $n = 0$ since $2^0 = 1 > 0$. Assume $2^k > k$, where $k \geq 0$. We show that $2^{k+1} > k + 1$. Observe that $2^{k+1} = 2 \cdot 2^k = 2^k + 2^k > k + 1$, since $2^k > k$ and $2^k \geq 1$ for $k \geq 0$.

47. The result is true for $n = 4$ since $4! = 24 > 2^4 = 16$. Assume that $k! > 2^k$, where $k \geq 4$. Then $(k + 1)! = (k + 1)k! > 2 \cdot k!$ (since $k \geq 4$). Since $k! > 2^k$ it follows that $2 \cdot k! > 2 \cdot 2^k = 2^{k+1}$.

49. The result is true for $n = 2$, since 2 is prime. Let $k(\geq 2)$ be an integer, and assume that all integers $2, 3, \ldots, k$ are either prime or can be written as a product of primes. Consider $k + 1$. If $k + 1$ is prime, the desired result follows. If $k + 1$ is not prime, then there exist integers a and b with $1 < a < k + 1$ and $1 < b < k + 1$ such that $k + 1 = ab$. By the induction hypothesis, each of a and b is prime or a product of primes. Hence $k + 1 = ab$ is a product of primes.

51. If two distinct lines intersect, then there is exactly one point of intersection, and $1 = 2(2 - 1)/2$. Assume that the maximum number of points of intersection of $k(\geq 2)$ distinct lines in the plane is $k(k - 1)/2$. Consider a collection of $k + 1$ distinct lines in the plane. Let l be one of them, and consider the remaining k lines. These k lines intersect in at most $k(k - 1)/2$ points of intersection. The line l can introduce at most k additional points of intersection. Therefore, all $k + 1$ lines produce at most $k(k - 1)/2 + k = k(k + 1)/2$ points of intersection.

53. By the product rule of differentiation from calculus,

$$f'(x) = xe^{-x}(-1) + e^{-x} = (-1)e^{-x}(x - 1),$$

and the result is true for $n = 1$. Assume for $k \geq 1$ that

$$f^{(k)}(x) = (-1)^k e^{-x}(x - k).$$

By the product rule, we have

$$\begin{aligned} f^{(k+1)}(x) &= (-1)^k[e^{-x} + (x - k)e^{-x}(-1)] \\ &= (-1)^{k+1}e^{-x}[x - (k + 1)]. \end{aligned}$$

55. For $n = 2$, the result is Theorem A.1(b). Assume the result for $n = k(\geq 2)$. Consider

$$\begin{aligned} \overline{A_1 \cap A_2 \cap \cdots \cap A_k \cap A_{k+1}} &= \overline{(A_1 \cap A_2 \cap \cdots \cap A_k) \cap A_{k+1}} \\ &= \overline{A_1 \cap A_2 \cap \cdots \cap A_k} \cup \overline{A_{k+1}} \qquad \text{[by Theorem A.1(b)]} \\ &= \overline{A_1} \cup \overline{A_2} \cup \cdots \cup \overline{A_k} \cup \overline{A_{k+1}} \qquad \text{[by the induction hypothesis].} \end{aligned}$$

Index

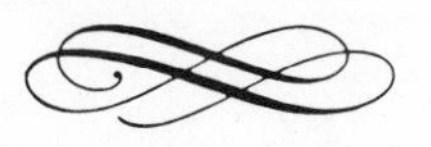